Beyond the Transistor:
133 Electronics Projects

No. 2887
No. $16.95

Beyond the Transistor:
133 Electronics Projects

RUFUS P. TURNER and BRINTON L. RUTHERFORD

TAB BOOKS Inc.
Blue Ridge Summit, PA 17214

SECOND EDITION
FIRST PRINTING

Copyright © 1987, 1986 by TAB BOOKS Inc.
Printed in the United States of America

Reproduction or publication of the content in any manner, without express permission of the publisher, is prohibited. No liability is assumed with respect to the use of the information herein.

Library of Congress Cataloging in Publication Data

Turner, Rufus P.
   Beyond the transistor : 133 electronics projects / by Rufus P. Turner and Brinton L. Rutherford.
     p. cm.
   Includes index.
   ISBN 0-8306-7887-5 : ISBN 0-8306-2887-8 (pbk.) :
   1. Electronics—Amateurs' manuals. I. Rutherford, Brinton L.
II.Title.
TK9965.T798   1987   87-19410
621.381—dc19   CIP

Questions regarding the content of this book should be addressed to:

   Reader Inquiry Branch
   Editorial Department
   TAB BOOKS Inc.
   P.O. Box 40
   Blue Ridge Summit, PA 17214

OTHER BOOKS IN THE
TAB HOBBY ELECTRONICS SERIES

This series of five newly updated and revised books provides an excellent blend of theory, skills, and projects that lead the novice gently into the exciting arena of electronics. The Series is also an extremely useful reference set for libraries and intermediate and advanced electronics hobbyists, as well as an ideal text.

The first book in the set, *Basic Electronics Course—2nd Edition*, consists of straight theory and lays a firm foundation upon which the Series then builds. Practical skills are developed in the second and third books, *How to Read Electronic Circuit Diagrams—2nd Edition* and *How to Test Almost Anything Electronic—2nd Edition*, which cover the two most important and fundamental skills necessary for successful electronics experimentation. The final two volumes in the Series, *44 Power Supplies for Your Electronic Projects* and *Beyond the Transistor: 133 Electronics Projects*, present useful hands-on projects that range from simple half-wave rectifiers to sophisticated semiconductor devices utilizing ICs.

Basic Electronics Course—2nd Edition
NORMAN H. CROWHURST
This book thoroughly explains the necessary fundamentals of electronics, such as electron flow, magnetic fields, resistance, voltage, and current.

How to Read Electronic Circuit Diagrams—2nd Edition
ROBERT M. BROWN, PAUL LAWRENCE, and JAMES A. WHITSON
Here is the ideal introduction for every hobbyist, student, or experimenter who wants to learn how to read schematic diagrams of electronic circuits. The book begins with a look at some common electronic components; then some simple electronic circuits and more complicated solid-state devices are covered.

How to Test Almost Everything Electronic—2nd Edition
JACK DARR and DELTON T. HORN
This book describes electronic tests and measurements—how to make them with all kinds of test equipment, and how to interpret the results. New sections cover logic probes and analyzers, using a frequency counter and capacitance meter, signal tracing a digital circuit, identifying unknown ICs, digital signal shaping, loading and power supply problems in digital circuits, and monitoring brief digital signals.

44 Power Supplies for Your Electronic Projects
ROBERT J. TRAISTER and JONATHAN L. MAYO
Electronics and computer hobbyists will not find a more practical book than this one. A quick, short, and thorough review of basic electronics is provided along with indispensable advice on laboratory techniques and how to locate and store components. The projects begin with simple circuits and progress to more complicated designs that include ICs and discrete components.

Contents

	Preface	ix
	Introduction	xi
	List of Projects	xiii
1	**Learning By Building**	**1**
	Tools • Components • Troubleshooting • Construction	
2	**FETs**	**11**
	Theory • Audio Preamplifier • Two-Stage Audio Amplifier • Untuned Crystal Oscillator • Tuned Crystal Oscillator • Phase Shift Audio Oscillator • Product Detector • Regenerative Receiver • Superregenerative Receiver • Electronic dc Voltmeter • Direct-Reading Capacitance Meter	
3	**MOSFETs**	**31**
	Theory • General Purpose Rf Amplifier • Ten-Meter Preamplifier • Wideband Instrument Amplifier • Source Follower • Q Multiplier • Wide-Range LC Checker • Interval Timer • Capacitance Relay • Touch-Plate Relay • Electronic Electroscope • Ultrasonic Pickup	
4	**Analog ICs**	**51**
	Theory • Audio Amplifier • Audio Amplifier with MOSFET Input • 2½-Watt Intercom • Low-Resistance Dc Milliammeter • Electronic Dc Millivoltmeter • Low-Pass Active Filter • High-Pass Active Filter • Combination Active Filter • Bandpass Active Filter • Active Notch Filter • Conventional Bandpass Active Filter • Signal Tracer • Code Practice Oscillator • Dual LED Flasher • Continuity Checker • AM Radio • Proximity Detector • Night Light • Electronic Noise Maker • Sound Machine	
5	**Digital ICs**	**82**
	Theory • Magnitude Comparator • Half Adder • Binary Counter • LSI Counter • Binary Logic Probe • Electronic Dice • Circle Chase • Melody Maker	
6	**CMOS ICs**	**101**
	Theory • Pendulum Clock • CMOS Logic Probe • Audio Thermometer • Random Number Generator • Egg Timer • Siren • Metronome • Frequency Counter • Music Box	

7 UJTs — 118

Theory • Pulse Generator • Pulse and Timing Generator • Free-Running Multivibrator • One-Shot Multivibrator • Relaxation Oscillator • Standard Frequency Oscillator • Cw Monitor • Metronome • Tone-Identified Signal System • Trigger for SCR

8 VCDs — 131

Theory • Voltage-Variable Capacitor • VCD-Tuned LC Circuits • Remotely-Controlled Tuned Circuit • Voltage-Tuned Rf Oscillator • Automatic FM Frequency Control • Frequency Modulators • VCD's Used to Fluctuate the Frequency of a Quartz Crystal • VCDs Used to Frequency Modulate Self-Excited Oscillators • Frequency Multipliers • Frequency Doubler • Frequency Tripler • Rf Harmonic Intensifier

9 Zener Diodes — 147

Theory • Simple Dc Voltage Regulator • Higher Voltage Dc Regulators • Multiple-Output Dc Voltage Regulator • Light-Duty Regulated Dc Supply • 5 V, 1.25 A Regulated Dc Supply • 18 V, 1 A Regulated Dc Supply • Voltage-Regulated Dual Dc Supply • Regulated Voltage Divider • Transistor Bias Regulator • Voltage Regulator for Tube Heater • Simple Ac Voltage Regulator • Automatic Volume Limiter • Dc Equipment Protector • Dc Voltage Standard

10 Emitting Diodes — 164

Theory • Three-Lead Tricolor LED • Two-Lead Tricolor LED • Seven-Segment Displays • Bar Graph Display • Infrared-Emitting Diode Transmissions • Infrared Emitting Diode Detection • Blinking LED • Tricolor Logic Probe • Decimal Counter • Simple Voltmeter • Dc Controlled Ac Lamp • Light Meter • Infrared Link • Fiberoptic Infrared Link

11 Diacs — 175

Theory • Amplitude Sensitive Switch • Static Dc Switch • Electrically Latched Relay • Latching Sensor Circuit • Dc Overload Circuit Breaker • Ac Overload Circuit Breaker • Phase-Controlled Trigger Circuit

12 Triacs — 185

Theory • Simple Triac Switch • General Purpose Controller • Controller Using Combination Thyristor • Dc-Controlled Solid-State Ac Relay • Sensitive Dc-Controlled Triac Switch • Motor Controls • Light Dimmers • Variable, Dual-Dc Power Supply • Automatic Equipment Power Switch

13 SCRs — 198

Theory • Basic SCR Switches • Light-Controlled SCR • SCR Light Dimmer • SCR Motor Control • Photoelectronic Burglar Alarm • Switch-Type Burglar Alarm • Photoelectric Garage-Door Opener • Variable Dc Power Supply • High-Voltage Dc Power Supply • Dc-to-Ac Inverter • Solid-State Timer

Index — 217

Preface

Since the first edition of this book appeared in 1976, the semiconductor industry has enjoyed considerable growth and expansion. Although integrated circuits were available in 1976, the level of integration and degree of functionality on a single chip of silicon were still low. Today, integrated circuits are applied in microwave ovens, toys, cars, televisions, radios, watches—and of course computers.

Therefore, it was necessary to update this book to reflect this trend in the industry. Powerful integrated circuits are readily available to the hobbyist, and this second edition presents many of these devices in practical circuits that you can use. Both analog and digital ICs are introduced.

The majority of the first edition is still intact. The previous material was reviewed for accuracy. Effort was made to identify obsolete part numbers. The result should be a useful and updated text to help you learn about all those semiconductor devices that reach beyond the bipolar transistor.

Introduction

This book is directed to the electronics hobbyist and experimenter who is already familiar with bipolar transistors and conventional diodes. The devices covered in this book are the result of improvements on the original bipolar transistor. Some of these devices have very specialized purposes. Others, like integrated circuits, threaten to overrun the transistor market because of their incredible flexibility and high functionality, yet compact design.

By studying this book and building the projects, you will learn how to use most of the semiconductor devices now available to the hobbyist. No rectifiers, conventional diodes, or bipolar transistors appear except where they serve as auxiliary components to the main circuit. Any student of electronics, whether formally enrolled in a course of study or informally learning in his spare time, can increase his ability to apply many different kinds of devices with this book.

The presentation is practical. Each chapter begins with a brief introduction, presents some background theory for the device type under study, and then illustrates the devices in practical, useful circuits. This format will help you to learn the theory of the device quickly, and it should help prepare you to apply the device in your own circuits. Each chapter also stands on its own. You can skip to any chapter that looks interesting and begin there.

The projects presented in this book were chosen because they seem to best exploit the advantages of the particular device. All the electronic components are listed within each circuit diagram. Secondary parts—sockets, chassis, enclosures, etc.—are not specified. The information in Chapter 1 and your previous project building experience with bipolar devices will allow you to decide on these details.

Even though you might be experienced in project building, I suggest you review at least Chapter 1 before proceeding to the projects. Chapter 1 presents some good background on breadboarding, finding components, and handling semiconductor devices.

List of Projects

Chapter 2
Project 1: Audio Preamplifier 14
Project 2: Two-Stage Audio Amplifier 15
Project 3: Untuned Crystal Oscillator 16
Project 4: Tuned Crystal Oscillator 17
Project 5: Phase-Shift Audio Oscillator 19
Project 6: Product Detector 20
Project 7: Regenerative Receiver 21
Project 8: Superregenerative Receiver 23
Project 9: Electronic dc Voltmeter 25
Project 10: Direct-Reading Capacitance Meter 27

Chapter 3
Project 11: General Purpose Rf Amplifier 34
Project 12: Ten-Meter Preamplifier 36
Project 13: Wideband Instrument Amplifier 38
Project 14: Source Follower 39
Project 15: Q Multiplier 41
Project 16: Wide-Range LC Checker 42
Project 17: Interval Timer 44
Project 18: Capacitance Relay 45
Project 19: Touch-Plate Relay 46
Project 20: Electronic Electroscope 48
Project 21: Ultrasonic Pickup 49

xiv Beyond the Transistor: 133 Electronics Projects

Chapter 4
Project 22: Audio Amplifier 55
Project 23: Audio Amplifier with MOSFET Input 57
Project 24: 2½-Watt Intercom 57
Project 25: Low-Resistance dc Milliammeter 59
Project 26: Electronic dc Millivoltmeter 60
Project 27: Low-Pass Active Filter 62
Project 28: High-Pass Active Filter 63
Project 29: Combination Active Filter 64
Project 30: Bandpass Active Filter 65
Project 31: Active Notch Filter 67
Project 32: Conventional Bandpass Active Filter 68
Project 33: Signal Tracer 69
Project 34: Code-Practice Oscillator 71
Project 35: Dual LED Flasher 72
Project 36: Continuity Checker 73
Project 37: AM Radio 74
Project 38: Proximity Detector 75
Projects 39 and 40: Night Light and Electronic Rooster 76
Project 41: Electronic Noise Maker 79
Project 42: Sound Machine 79

Chapter 5
Project 43: Magnitude Comparator 87
Project 44: Half Adder 89
Project 45: Full Adder 90
Project 46: Binary Counter 92
Project 47: LSI Counter 93
Project 48: Binary Logic Probe 94
Project 49: Electronic Dice 95
Project 50: Circle Chase 97
Project 51: Melody Maker 99

Chapter 6
Project 52: Pendulum Clock 103
Project 53: CMOS Logic Probe 105
Project 54: Audio Thermometer 106
Project 55: Random Number Generator 107
Project 56: Egg Timer 107
Project 57: Siren 110
Project 58: Metronome 111
Project 59: Frequency Counter 113
Project 60: Music Box 115

Chapter 7
Project 61: Pulse Generator 120
Project 62: Pulse and Timing Generator 121
Project 63: Free-Running Multivibrator 122
Project 64: One-Shot Multivibrator 123
Project 65: Relaxation Oscillator 124
Project 66: Standard-Frequency Oscillator 125
Project 67: Cw Monitor 126
Project 68: Metronome 128
Project 69: Tone-Identified Signal System 129
Project 70: Trigger for SCR 130

Chapter 8
Project 71: Voltage-Variable Capacitor 133
Projects 72-74: VCD-Tuned LC Circuits 135
Project 75: Remotely Controlled Tuned Circuit 137
Project 76: Voltage-Tuned Rf Oscillator 138
Project 77: Automatic FM Frequency Control 139
Project 78: VCDs Used to Frequency Modulate Self-Excited π Oscillators 141
Project 79: VCDs Used to Fluctuate the Frequency of a Quartz Crystal 142
Project 80: Frequency Doubler 143
Project 81: Frequency Tripler 144
Project 82: Rf Harmonic Intensifier 145

Chapter 9
Project 83: Simple dc Voltage Regulator 149
Project 84: Higher Voltage dc Regulators 150
Project 85: Multiple-Output dc Voltage Regulator 151
Project 86: Light-Duty Regulated dc Supply 152
Project 87: 5 V, 1.25 Amp Regulated dc Supply 153
Project 88: 18 V, 1 Amp Regulated dc Supply 155
Project 89: Voltage-Regulated Dual dc Supply 156
Project 90: Regulated Voltage Divider 157
Project 91: Transistor-Bias Regulator 158
Project 92: Voltage Regulator for Tube Heater 159
Project 93: Simple ac Voltage Regulator 160
Project 94: Automatic Volume Limiter 161
Project 95: Dc-Equipment Protector 162
Project 96: Dc Voltage Standard—Single Stage 162
Project 97: Dc Voltage Standard—Two Stage 163

Chapter 10
Project 98: Blinking LED 167
Project 99: Tricolor Logic Probe 167
Project 100: Decimal Counter 168
Project 101: Simple Voltmeter 169
Project 102: Dc Controlled ac Lamp 171
Project 103: Light Meter 172
Project 104: Infrared Link 173
Project 105: Fiberoptic Infrared Link 174

Chapter 11
Project 106: Amplitude-Sensitive Switch 177
Project 107: Static dc Switch 178
Project 108: Electrically Latched Relay 179
Project 109: Latching Sensor Circuit 180
Project 110: Dc Overload Circuit Breaker 181
Project 111: Ac Overload Circuit Breaker 182
Project 112: Phase-Controlled Trigger Circuit 183

Chapter 12
Project 113: Simple Triac Switch 187
Project 114: General-Purpose Controller 189
Project 115: Controller Using Combination Thyristor 190
Project 116: Dc-Controlled Solid-State ac Relay 191
Project 117: Sensitive dc-Controlled Triac Switch 192
Projects 118 and 119: Two Improved Light Dimmers 194
Project 120: Variable Dual dc Power Supply 195
Project 121: Automatic Equipment Power Switch 197

Chapter 13
Projects 122 and 123: Basic SCR Switches—Ac and Dc 200
Project 124: Light-Controlled SCR 202
Project 125: SCR Light Dimmer 203
Project 126: SCR Motor Control 204
Project 127: Photoelectric Burglar Alarm 205
Project 128: Switch-Type Burglar Alarm 206
Project 129: Photoelectric Garage-Door Opener 207
Project 130: Variable dc Power Supply 209
Project 131: High Voltage Variable dc Power Supply 210
Project 132: Dc-to-ac Inverter 212
Project 133: Solid-State Timer 214

1
Learning by Building

There is no substitute for experience. While you can learn facts by reading, rarely can you understand concepts without doing. Gaining experience by actually playing with electronic components in a circuit pays high dividends. The goal of this book is to present families of components and circuits using members of that family to illustrate why they are useful in certain situations, how they can be used, and what precautions to take. If you build the projects in this book and experiment with them until you understand the component under study, you will have gained an incredible wealth of electronic knowledge.

Before rushing off to begin building, let me present some project building basics. To efficiently and easily build electronics projects, there are a few things that you will want to prepare beforehand. First, you will need a few tools. Not many tools are needed, but a few key tools will make your adventure less frustrating. You will also need a supply of components. Next, a method of building a circuit that allows easy changing and rearranging of the components is a necessity. Finally, you need some ideas about what to do when the circuit you have just put together does not function.

If you build a temporary circuit that you want to make more permanent, then you will be interested in some physical construction details. This is the last item in this chapter. From here, all that remains is to begin learning by building. Each chapter stands on its own, so you can start wherever you like.

TOOLS

Everybody knows that the right tool for the job always makes that job go easier. Electronics is no exception. There are a few tools that will make your experimentation easier and more enjoyable. Most of these you probably already have, and the ones that you do not have are not overly expensive.

Work Area

The first prerequisite to efficient project building and experimentation is a place to work. It is important to be able to set up a circuit and not have to disassemble everything when you are done for the day. Many times you will want, or need, to leave a circuit up for quite a while. You also need a work place where you can store your components and other tools so that everything is very handy.

Another need of the work area is good lighting. Some of the components are small and when the breadboard starts getting a little crowded it is very helpful to have good light.

Hand Tools

A diagonal cutting pliers, a needle-nose pliers, a small straight blade screwdriver, and a small Phillips screwdriver are the major hand tools that you will find useful. I also routinely use a small X-Acto knife and about a half dozen or so 6-inch leads with alligator clips on each end. These last items are so handy that once you have some you will always need more. You can make them very easily, or simply spend a couple dollars for some.

A good soldering iron is a necessity. One with a fine tip and a 25-watt element will do nicely. You will also need some fine rosin core solder. I use 60 percent tin, 40 percent lead solder mixture in the .032-inch-diameter size. For working with the pads on a circuit board, this size is excellent.

Two other specialized hand tools that I use are an IC extractor and an IC inserter/pin straightener. I would not describe these tools as a necessity. However, they do make things easier. After you have pushed the pins of an IC into your thumb while pulling it out of the breadboard, or bent over the V+ pin again on your last 556 timer so that it breaks off when you straighten it out, you will appreciate the value of these tools.

Breadboard

A breadboard allows the experimenter to build a circuit without having to solder anything in place. It holds all the leads very firmly in place, but it is not a permanent connection. Once you have the circuit working the way you want it to, you can easily swap out components for experimentation or transfer it to a circuit board. There is no substitute for a good breadboard. Simple models can be purchased from Radio Shack, but some very sophisticated ones are available from AP Products, which are available through DIGI-KEY's catalog. The catalog is free for the asking.

DIGI-KEY CORPORATION
701 Brooks Ave. South
PO Box 677
Thief River Falls, MN 56701

A quality breadboard will have a breadboarding area for mounting components (including DIP ICs), a power supply (+5 V and +12 V), a function generator (variable frequency square and sine wave), and some logic switches and LEDs. To effectively use the breadboard, you will need a large number of wires in various lengths with the insulation removed from both ends. These are used to connect various points on the breadboard.

Test Equipment

Several simple test equipment projects appear in this book. There are continuity testers, logic probes, various meters, signal generators, and a frequency counter. These instruments do not approach the quality and accuracy of commercial products. They will often suffice for simple troubleshooting and experimenting with the projects in this book, however.

In addition to the test instrument projects, you might want to have a quality multimeter. Again, both Radio Shack and DIGI-KEY are sources for a multimeter. The ability to accurately measure resistance, voltage, and current is helpful, but it is not something that is needed all the time. If you do not already have a meter, wait until you can afford a fairly good one.

Another piece of test equipment that is not used very often, but that can easily earn its keep for serious electronics work, is the oscilloscope. This is an expensive piece of equipment, but it is very versatile.

Safety

The voltages used by the majority of the circuits in this book are harmless. However, any of the circuits that involve house current (120 Vac) can kill you. A finger across 120 Vac and ground will give you a nasty shock, and 120 Vac across your chest will stop your heart. It is imperative that proper respect is given to electricity. Often, the low voltages used in TTL and CMOS IC circuits can lull us into a sense of security. Try to always maintain an attitude that a lethal voltage might be present. This "better-safe-than-sorry attitude" can keep you alert to potential danger.

COMPONENTS

The components used in this book are readily available from electronic parts retailers and mail-order parts houses. In addition, after you have been building projects for awhile, you will begin to collect quite an assortment of parts yourself. Other hobbyist that you know also become sources for parts.

It is also important to handle electronics components with care. Although they can be surprisingly rugged, it is quite possible to destroy them. They can also be damaged by careless or improper mounting on a breadboard or circuit board.

Finding Components

Anyone who has built electronic projects has experienced the frustration of getting the necessary parts. I suggest that you begin to collect as many catalogs from as many parts suppliers as possible. The following list should get you started.

ALPHA ELECTRONIC LABORATORIES
2302 Oaland Gravel Road
Columbia, MO 65201

DICK SMITH ELECTRONICS
PO Box 8021
Redwood City, CA 94063
1-800-332-5373

DIGI-KEY CORPORATION
701 Brooks Ave. South
PO Box 677
Thief River Falls, MN 56701
1-800-344-4539

ELECTRONIC PARTS SUPPLY, INC.
PO Box 29321
Oakland, CA 94604
1-800-227-0104

JAMECO ELECTRONICS
1355 Shoreway Road
Belmont, CA 94002

MCM ELECTRONICS
PO Box 650
Dayton, OH 45449
1-800-551-1522

R&D ELECTRONICS
1202H Pine Island Road
Cape Coral, FL 33909
1-813-772-1441

R&D ELECTRONIC SUPPLY
100 E. Orangethorpe Ave.
Anaheim, CA 92801
1-714-773-0240

Radio Shack is also a very useful source of the more common items. The Radio Shack catalog is a useful addition to your resource materials.

Substituting Components

Also, be alert to the possibility of substituting one component for another. Often your junkbox will have a part that will work in place of the part called out in the schematic. Strive to improve your abilities at circuit analysis. If you can determine whether the part in question is a noncritical component, then probably any substitute that is close to the value or the parameters of the device will work. One of the best ways to discover critical components is to switch different parts into a circuit once it is working. This way you learn why certain components are used and what effect they have on the circuit.

Substitution books and Replacement Guides are also helpful in project building. Several good ones are available from TAB Books.

The Master Semiconductor Replacement Handbook—Listed by Industry Standard Number
TAB Book #1470, $25.95

The Master Semiconductor Replacement Handbook—Listed by Manufacturer's Number
TAB Book #1471, $25.95

Tower's International Transistor Selector—3rd Edition
TAB Book #1416, $19.95

Tower's International Digital IC Selector
TAB Book #1616, $19.95

Linear IC Handbook by Michael S. Morely
TAB Book #2672, $48.50

Handling Components

Modern solid-state devices are remarkably well built, considering their smallness and complexity, and many of them are surprisingly inexpensive. Moreover, their performance is consistent. However, their electrical and mechanical ruggedness, easy availability, and reproducible performance should not be taken for granted. Careless handling, installation, and operation of these devices and misguided application of them can still cause damage or poor performance.

The following paragraphs offer guidance in working with semiconductor devices, and these hints and precautions are offered for the benefit of all readers. Observance of good engineering practice here will pay off in increased success with semiconductor devices.

This material applies directly or indirectly to all of the circuits in this book. By presenting it in this one place, I avoid taking space and time to repeat hints and precautions with individual circuits.

Avoid Rough Handling. This includes dropping, hammering, vibrating, or forcibly squeezing or tensing the device; pulling, twisting, or repeatedly flexing its pigtails or lugs; and mounting the device under severe pressure or tension.

Install the Semiconductor Last. In this way, the device will not be subjected to repeated heatings caused by soldering in the circuit. It is important to check the circuit wiring thoroughly before installing the semiconductor device to insure that the device will not be heated by unsoldering and resoldering to correct mistakes in the wiring. When the device is a standard MOSFET, keep the short-circuiting ring (that comes attached to this device) in place until the MOSFET is completely installed in the completed circuit; then remove it. The gate-protected MOSFET needs no such protecting ring.

Remove the Semiconductor First. When dismantling a circuit, remove the semiconductor device first. This procedure, which protects the device from heat caused by unsoldering, is the opposite of that recommended when building the circuit.

Semiconductor Mounting. All methods of device mounting are acceptable as long as the mounting is solid. These include use of socket, soldering or welding directly into circuit, use of clips, use of mounting screws, and use of terminal studs. The pigtails or lugs of the mounted device must be under no stress.

Soldering. To prevent internal damage to the semiconductor device, use a suitable heat sink in the regular manner when soldering the device into the circuit. Keep the heat sink in place for a reasonable time after the joint has cooled.

Keep Leads Straight. If a semiconductor device is to be inserted into a socket, straighten its pigtails or lugs beforehand. To drive the device home into the socket, push firmly, but gently and straight down, on the top of the case. When leads must be bent for a particular installation, avoid a sharp right-angle bend, since such a bend tends to break easily. Use care when removing a semiconductor device from its socket.

Hot Case. The metal case of some semiconductors is electrically "hot;" that is, the case is internally connected to one of the electrodes of the device. If the manufacturer's literature or the circuit diagram and/or the text in this book indicates this condition, keep the case free from contact with other components, metal chassis, and wiring.

Wiring And Isolation. Use the shortest and most direct leads practicable; this will minimize stray pickup, undesired coupling, and undesired feedback. When long leads are unavoidable, lead dress and adequate separation are important. With high-grain devices, such as the FET, MOSFET, and IC, the input and output circuits sometimes must be shielded from each other, especially if they employ inductors or transformers. In all rf and high-gain af circuits, shield and bypass all susceptible parts of the circuit in the same way that tube circuits are safeguarded.

Grounds. In some of the circuit diagrams, a dashed line runs to the ground symbol. This means that the connection to chassis or to earth is optional and depends upon how the circuit will be used by the reader. When, instead, a solid line runs to the ground symbol, the connection must be made.

Connect Power Last. Connect the power supply last, whether it is ac or dc, and then only after the circuit wiring has been double checked and verified as correct. In an experimental application—where voltages are not specified as they are in this book—start with a low voltage and gradually increase it.

Type of Power Supply. In many of the circuits, batteries are shown for dc supply, as a matter of simplicity; however, a well-filtered transformer rectifier type of supply also may be used.

Complexity of Power Supply. Some IC circuits function best with a dual power supply and accordingly are shown with two batteries; others get along with a single battery. Readers who already are expert with ICs may have favorite schemes for converting the two-battery circuit to single battery; however, it is best for all others to wire and test the circuit first as it is given in the book, and then to experiment later.

Use Specified Voltages. Employ the voltages specified in the circuit diagrams. Although a variation of a few percent, plus or minus, should not drastically affect operation of a circuit, a very large change in voltage can alter performance markedly from that reported in the text.

Dc Polarity. Reversing the dc supply voltage can damage some semiconductor devices and cause others to switch off. Carefully observe polarity.

Avoid Excessive Supply. Do not subject the semiconductor device to excessive currents or voltages. Be guided by the values given in the circuit diagrams and in the text. Never allow the combined current or voltage (that is, dc plus peak ac) to exceed the maximum value given in the manufacturer's literature or warned against in the text.

Avoid Heat. Protect the semiconductor device from excessive heat, either external or internal. Where a heat sink is specified, it must be used. Keep the device clear of hot tubes, rectifiers, and other such components.

Avoid External Fields. The semiconductor device and its circuit must be protected from strong, external magnetic fields. Common sources of such fields are transformers, chokes, motors, generators, relays, circuit breakers, and loudspeakers.

Avoid Overdriving. Excessive signal amplitude can degrade the performance of some semiconductor devices and in some instances may even damage them. The maximum signal voltage permissible in a susceptible circuit is given in the circuit diagram or in the text, and should not be exceeded.

TROUBLESHOOTING

Despite your best efforts, sometimes a circuit that you breadboard does not work. The challenge then becomes "Why doesn't it work?" Another problem that can be encountered is a circuit that was long ago breadboarded and subsequently transferred to a pc board successfully. Then, after much use, one day it quits. The techniques for finding the trouble is referred to as troubleshooting. While the principles are mostly the same, these two situations require a slightly different mode of attack.

Breadboard Troubleshooting

When I transfer a new design from paper to the breadboard, I keep two things in mind. First, I probably neglected something on the paper design—a crucial connection, a wrong voltage, or a wrong assumption. Second, when I breadboard the circuit, I will probably make an error in connecting the wires. This mental attitude has me expecting the worst. When a circuit works right the first time, I am always pleasantly surprised. When it doesn't, I am already expecting it.

Therefore, the first step in fixing a faulty design is to review the paper design for errors. Next check all the breadboarded connections. Is power connected to the circuit? Is ground connected? Are the correct chips or other devices in the circuit? The answers to these questions can be embarrassing, but they will probably get results.

When I breadboarded the Decimal Counter project in Chapter 10, I made a crucial error. I put a 7447 IC into the circuit with a common cathode seven-segment display. Because the 7447 drives a common anode display, not a common cathode display, the circuit did not work. I scratched my head over this circuit for two days. I checked the paper design. I checked the connections on the breadboard. Finally, I traced the signal through the circuit, and realized my error. A simple substitution of a 7448 for the 7447 solved the problem.

This incident also points out the third step in troubleshooting a circuit. Study the circuit and determine what kind of signal should be at specific points in the circuit. Then check to see if the expected signal is present. For instance, when tracing the Decimal Counter circuit, I checked to see if the outputs from the counter were stepping through the BCD count properly. I determined with a logic probe that the chip was counting. This information exonerated the 7490 counter as the problem. Next I applied a low to the lamp test pin on the 7447. None of the segments on the display lit up. I still did not know whether the problem was with the driver or the display from this information, so I made another check.

I decided to put a voltage through a current-limiting resistor to one of the segments of the display. The segment lit up. Now I knew the problem was with the display driver. As soon as I pulled the chip off the board, I realized the mistake. I substituted the 7448 display driver, and the circuit was up and running.

This three-step process will find all the circuit bugs that you will ever have. First check the design; then check the connections, and finally, determine what devices are functioning and which are not. By collecting test information, you can work from the known and deduce where the error must be located.

Working Projects

When a project that has been working previously suddenly quits, the process is a little bit simpler than troubleshooting a new design. Because the project once worked, you know that the design worked. Something has changed in the circuit that was once OK.

The first step is to give the circuit a visual inspection. Has a wire broken? Are there any burnt components? Is the battery connected or the fuse blown? If a visual inspection does not turn up a solution, then you will have to work your way through the circuit as discussed in step three in the previous section.

Although the troubleshooting process can be rather frustrating, practice will make perfect. Actually it is quite thrilling to have solved a problem in a circuit. It is a very satisfying experience. If you spend much time working on electronics projects, you will certainly get plenty of practice troubleshooting. It goes with the territory.

CONSTRUCTION

When a circuit has been breadboarded and tinkered with for a while, you may want to make it a permanent project. Doing so is not very hard. A few guidelines should provide you with the help you will need.

Circuit Boards

There are three main ways that a circuit can be made permanent: perfboard, universal circuit board, and etched circuit boards. Perfboard is an option, but I find the result always a little sloppy. Etched circuit boards are very nice but there is a lot of effort in making a board for only one project. For the projects in this book, and most projects of a size similar to the ones in this book, the universal boards are the best choice.

Universal circuit boards are etched in standard commonly used patterns. They will accommodate both transistor and IC circuits. Most of them have a number of pads connected together for multiple connections to a point or IC pin. Connections between points on the board are made in the same way as on the breadboard—with short pieces of insulated connecting wire. In this case, however, instead of just pushing the wire into the breadboard, the wire is soldered to the pads on the universal board.

These boards come in various shapes and sizes. Some of the larger boards will accommodate five ICs on a single board with adequate room for the point-to-point wiring and the power and ground bus. They are available from Radio Shack and some of the other sources listed earlier.

IC Sockets

I highly recommend the use of IC sockets. By using sockets to mount ICs on a circuit board, you can always salvage the chip out of one project to use in another. It also simplifies soldering. You do not need to concern yourself with overheating the chip. Should a chip fail in one of the projects that you have built on a circuit board, replacement is also very easy. Sockets are very inexpensive, and they can be bought inexpensively in large quantities. If you build a lot of projects, it just makes sense to use sockets.

Enclosures

A nice enclosure makes a nice finishing touch to a project. Commercially available enclosures are available, and they will work just fine. They are also fairly inexpensive. But they are not the only source of enclosures. I have seen projects mounted inside of cassette cases, panty hose containers, pen barrels, egg cartons, and other exotic containers. Before spending the money for an enclosure, you might want to look around your bench, the basement, and the area under the kitchen sink to see if anything suitable catches your eye. It is also easy to build your own enclosures. Wood, Plexiglas, and light-gauge aluminum are all possibilities.

When designing an enclosure, give some thought to the placement of the controls, the display, and the labeling of them. It is important for the enclosure to enhance the utility of the project that you have built. Lay out the front panel so that it is pleasing to the eye as well as giving easy access to the controls.

Documentation

Documentation is often an item that hobbyists forget about. It is a good idea to document a project that you build with a schematic and a general discussion of how it functions. If the circuit fails for some reason at a later date, it will be much easier to trace down the problem with this information. When a project breaks years after you have built it, it is very difficult to remember exactly how it was designed.

Usually, you will end up working out the schematic from the circuit board and analyzing the circuit operation. A little documentation can save many lost hours on this task. When it is fresh in your mind, it is an easy task to neatly draw out the schematic and jot down some notes on circuit function.

2
FETs

The field-effect transistor (FET) exhibits high input impedance (in the megohms) and accordingly presents negligible loading to a signal source or preceding stage. In this respect, the FET, unlike the conventional bipolar transistor, behaves more like a vacuum tube than a semiconductor device. FET transconductance is high (1000 to 12,000 micromhos, depending upon make and model) and maximum operating frequency likewise is high (up to 500 MHz for some types). This device is useful in all kinds of electronic circuits and can directly replace the tube in some of them.

The kind of FET featured in this chapter is the junction type (JFET). It is composed of junctions in a silicon chip, but it is quite different from a junction-type bipolar transistor, as is explained below.

THEORY

Figure 2-1A shows the cross section of an FET. A manufactured unit is somewhat more complex in structure, but the illustration is functionally correct and will demonstrate the basic structure and behavior of the device. In this FET, p-regions are processed into opposite faces of an n-type chip and are connected together to form the control electrode (called the gate, G) which corresponds to the grid of a tube or the base of a bipolar transistor.

An ohmic connection is made at each of the two ends of the chip. One of these is termed the drain (D) and corresponds to the plate of a tube or the collector of a bipolar transistor, and the other is termed the source (S) and corresponds to the cathode of a tube or the emitter of a bipolar transistor. The region extending internally from drain to source and passing between the two halves of the gate electrode is termed the channel.

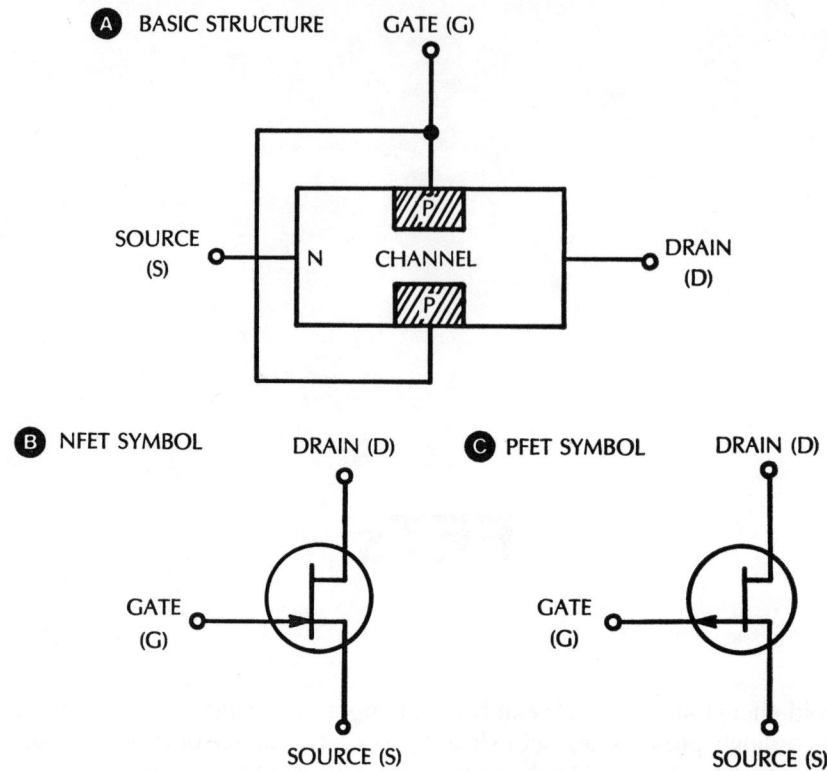

Fig. 2-1. Details of a junction FET with schematic symbols.

Although Fig. 2-1A shows p-type gate electrodes in an n-type chip, the opposite arrangement also is possible. N-type gate electrodes can be processed into a p-type chip. An FET employing an n-type chip is termed an n-channel FET or an NFET, and its circuit symbol appears in Fig. 2-1B. An FET employing a p-type chip is termed a p-channel FET or a PFET, and its circuit symbol appears in Fig. 2-1C.

Figure 2-2 illustrates FET operation. A dc supply (VDD) provides an operating voltage (VDS) between drain and source, with the drain positive and the source negative. A second dc supply (VGG) provides a bias voltage (VGS) between gate and source, with the gate negative and the source positive. In Fig. 2-2A, the gate voltage is zero, and under this condition, a maximum value of drain current (I_D) flows from supply VDD, through the channel, and back to VDD. In Fig. 2-2B, the gate voltage has a low negative value.

The application of this voltage causes a depletion region—an area in which no current carriers exist—to appear around each of the p-regions of the gate electrode (shown by the dotted lines). These layers penetrate deep into the chip and narrow the channel, thereby reducing drain current I_D. When the gate voltage is increased to a sufficiently high negative value, the depletion layers meet and block the channel, cutting off the flow of drain current.

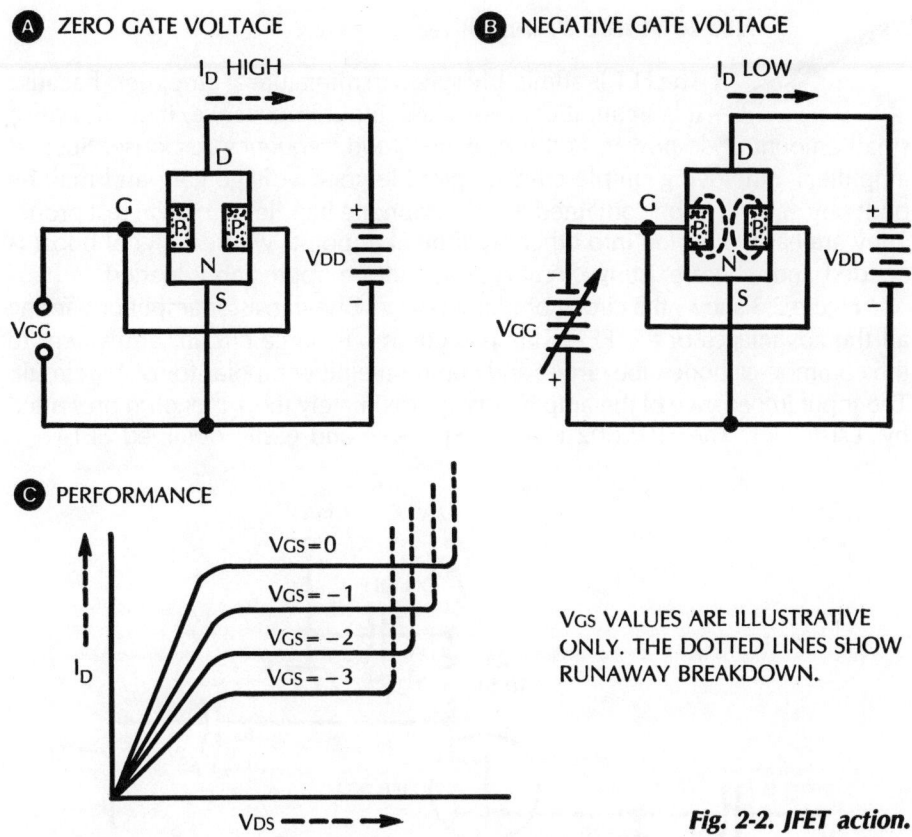

Fig. 2-2. JFET action.

Between these limits of maximum current and cutoff, the drain current may be set to any intermediate value by appropriately setting the gate voltage. Because the drain current is controlled by gate voltage, the FET possesses transconductance and is a good amplifier. And because the control (gate) electrode is a reverse biased pn junction, any control current is negligible; that is, the input resistance of the device is very high. Figure 2-2C shows FET performance in a family of curves resembling those for a pentode tube. An NFET is shown in Fig. 2-2; for a PFET, reverse the polarity of V_{DD} and V_{GG}. The performance curves will remain substantially the same as in Fig. 2-2C.

No. 1: Audio Preamplifier

The FET is admirably suited to miniature af amplifiers because it is small, it possesses high input impedance, it needs only a small amount of dc power, and it provides good frequency response. Such af amplifiers, employing simple circuits, provide good voltage gain and may be built tiny enough to be contained in a microphone handle or in an af test probe. They are easily inserted into other equipment at points where a signal boost is needed and where existing circuitry must not be appreciably loaded.

Figure 2-3 shows the circuit of a single-stage, one-transistor amplifier offering all the advantages of the FET. This is a common-source circuit, equivalent to the common-cathode tube circuit and common-emitter bipolar-transistor circuit. The input impedance of the amplifier is approximately the 1 megohm presented by resistor R1. The HEP 802 is an inexpensive and easily obtained FET.

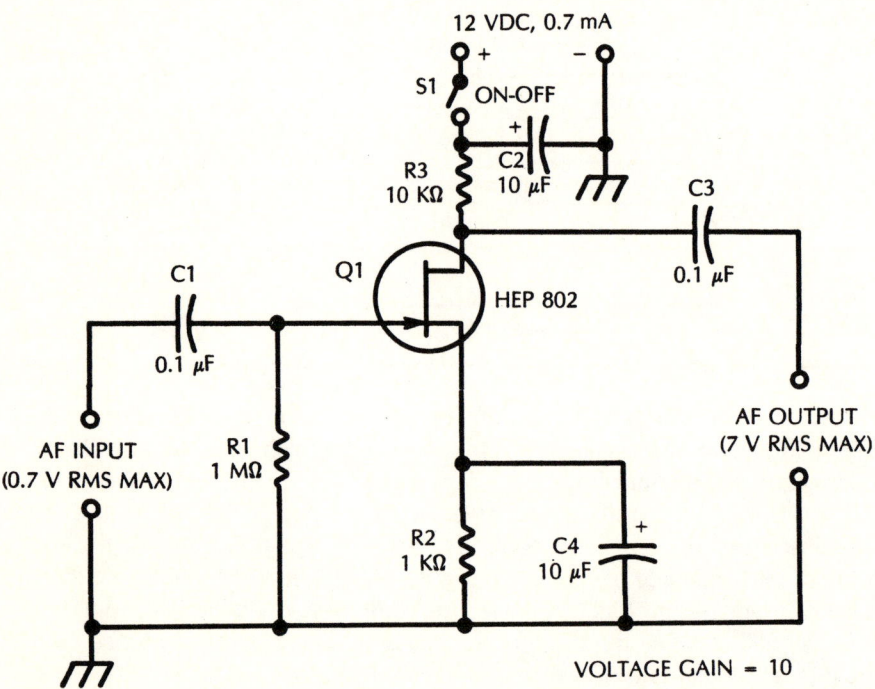

Fig. 2-3. Single-stage af preamplifier.

Voltage gain of the amplifier is 10. The maximum input-signal amplitude before output-signal peak clipping is approximately 0.7 volts rms, and the corresponding output-voltage amplitude is 7 volts rms. At full operation, the circuit draws 0.7 mA from the 12-Vdc supply. With an individual FET, the input-signal voltage, output-signal voltage, and dc operating current may differ somewhat from the values given above. Between 100 Hz and 25 kHz, the frequency response of the circuit is within 1 dB of the 1000 Hz reference.

All resistors are ¼- or ½-watt. Electrolytic capacitors C2 and C4 are 35-volt units, and capacitors C1 and C3 may be any convenient low-voltage units. Any type of battery or other dc power supply can be used; the amplifier may even be sun powered by two 5 volt silicon solar modules connected in series. If desired, continuously variable gain control may be obtained by substituting a 1 MΩ potentiometer for resistor R1.

This circuit is suitable as a preamplifier or as a main amplifier in all applications requiring a 20 dB signal boost throughout the audio spectrum. The high input impedance and medium output impedance will satisfy most requirements.

No. 2: Two-Stage Audio Amplifier

Figure 2-4 shows the circuit of a two-stage FET amplifier consisting of two identical RC-coupled stages of the type discussed in the previous section. This unit will give a substantial boost (40 dB) to a small af signal, and can be used either alone or as a stage in other equipment. The HEP 802 transistors are low-priced and readily available. The input impedance of the amplifier is approximately 1 megohm, the resistance of the input-stage resistor, R1.

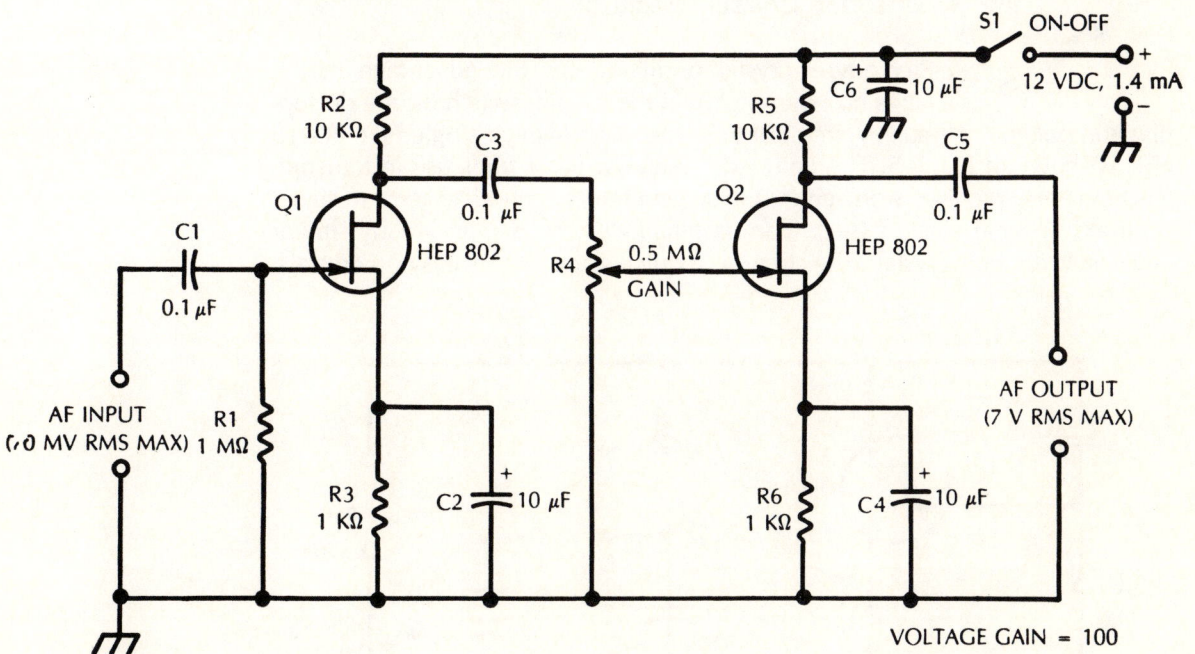

Fig. 2-4.

Overall voltage gain of the amplifier is 100, but this figure may vary somewhat—up or down—with individual FETs. The maximum input-signal amplitude before output-signal peak clipping is 70 mV rms, and the corresponding output-signal amplitude is 7 volts rms. At full operation, the circuit draws approximately 1.4 mA from the 12-Vdc supply, but this current may vary a small amount with individual FETs. (No need was found for a decoupling filter between stages. Such a filter will decrease the current of one stage.) The frequency response of the amplifier is flat within ± 1 dB of the 1 kHz level from 100 Hz to better than 20 kHz.

All resistors in the circuit are ¼- or ½-watt. Electrolytic capacitors C2, C4, and C6 are 25-volt units, and capacitors C1, C3, and C5 may be any convenient low-voltage units. The amplifier can be operated from a self-contained 12-volt battery or from an external battery or line-operated power supply.

Since the input stage runs "wide open," there may be some tendency to pick up hum and noise unless this stage and the input leads are well shielded. In stubborn cases, R1 may be reduced to 0.47 MΩ. Where the amplifier must introduce only small loading of the signal source, R1 may be increased to as a high a value as 22 MΩ, provided the input stage is very well shielded. However, resistance higher than this value tends to approach the FET junction resistance.

No. 3: Untuned Crystal Oscillator

A Pierce-type crystal oscillator has the advantage that it requires no tuning; plug in the crystal, switch on the dc supply, and obtain rf output. A circuit of this type, employing a single HEP F0015 FET, is shown in Fig. 2-5. The untuned crystal oscillator finds use in transmitters, markers, receiver front ends, clock generators, crystal testers, rf signal generators, signal spotters (secondary frequency standards), and many similar devices. With most crystals, the circuit is a quick starter that is easy on crystals.

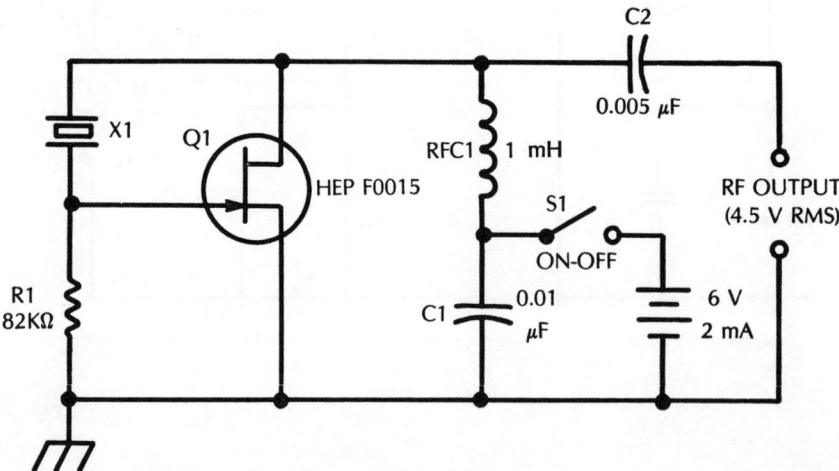

Fig. 2-5.

The untuned oscillator draws approximately 2 mA from the 6-Vdc supply. At this supply voltage, the open-circuit rf output voltage is approximately 4½ volts rms. A dc supply voltage of up to 12 volts can be used with proportionately higher rf output.

To determine if the oscillator is operating, close switch S1 and connect an rf voltmeter to the rf Output terminals. (A high-resistance dc voltmeter shunted by a general-purpose germanium diode will suffice if an electronic voltmeter with rf probe is not available.) Deflection of the meter indicates oscillation. Alternatively, the oscillator may be connected to the antenna and ground terminals of a CW receiver which may be tuned through the crystal frequency to detect oscillation.

To prevent erroneous operation, the user must remember that the Pierce oscillator operates at the labeled frequency of a crystal only when the crystal is a fundamental-frequency cut. When overtone crystals are used, the oscillator output will not be the labeled frequency, but the lower frequency determined by the crystal dimensions. For operation at the labeled frequency of an overtone crystal, the oscillator must be of the tuned type.

No. 4: Tuned Crystal Oscillator

Figure 2-6 shows the circuit of a general-purpose crystal oscillator which will operate with all types of crystals. The FET is an inexpensive and readily available HEP 801. The circuit is tuned by means of the screwdriver-adjusted slug in inductor L1. This oscillator is readily adapted to communications, instrumentation, and control applications. It may even be used as a flea-powered transmitter (dc power input, 12 mW) for communications or radio model control.

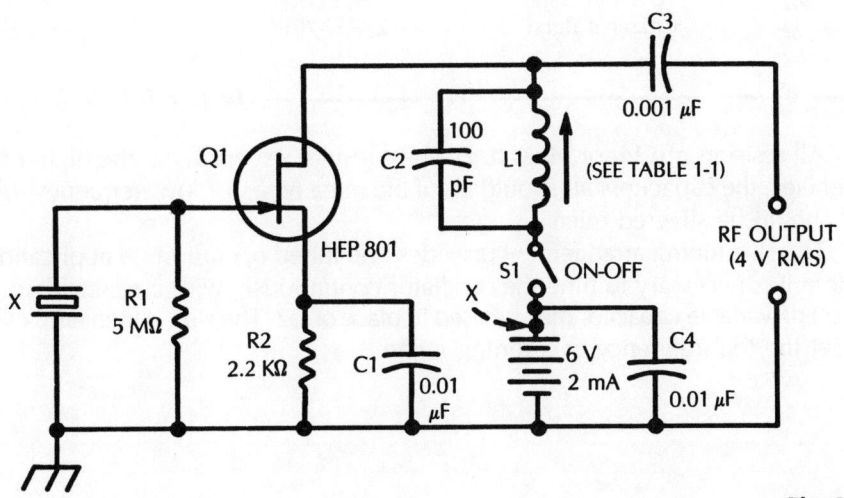

Fig. 2-6.

When the resonant circuit, L1-C1, is tuned to the crystal frequency, the oscillator draws approximately 2 mA from the 6-volt dc supply. The corresponding open-circuit rf output voltage is approximately 4 volts rms (dc and rf values both will vary somewhat with individual FETs). The drain current is lower on 100 kHz than on other bands, owing to the resistance of the inductor used for that frequency.

Table 2-1 lists commercial, slug-tuned inductors (L1) that can be used in this oscillator. Inductances have been chosen for the 100 kHz standard frequency, five ham radio bands, and the 27 MHz citizens' band; however, an appreciable inductance range is covered by adjustment of the slug of each inductor, and a wider frequency range than each band indicated in the table can be obtained with each inductor.

The oscillator may be tuned to a crystal frequency by adjusting the slug of the corresponding inductor (L1) for maximum deflection of an rf voltmeter connected to the rf Output terminals. (If an electronic voltmeter with an rf probe is not available, a high-resistance dc voltmeter shunted by a general-purpose germanium diode will suffice.) Alternatively, the oscillator may be tuned with a 0-5 mA dc milliammeter inserted temporarily at point X: Adjust the slug of inductor L1 for deepest dip in the meter reading.

FREQUENCY	J.W. MILLER COIL NO.
100 kHz	23A332RPC
160-Meter Band	23A825RPC
80-Meter Band	23A225RPC
40-Meter Band	23A476RPC
20-Meter Band	23A156RPC
10-Meter Band	23A337RPC
Citizen's Band	23A337RPC

Table 2-1. Coil Data.

All resistors are ¼- or ½-watt. For best results, especially at the higher frequencies, the capacitors all should be of the mica type; for low frequency drift, C2 should be silvered mica.

The slug tuning arrangement provides fine-tuned operation. In applications where it is necessary to tune the oscillator continuously with a resettable dial, a 100 pF variable capacitor may be used in place of C2. The slug is then employed to set the top frequency of a tuning range.

No. 5: Phase-Shift Audio Oscillator

The phase-shift oscillator is a relatively simple resistance-capacitance tuned circuit which is prized for its clean output signal (low-distortion sine wave). The field-effect transistor is well suited to this circuit, since the high input impedance of this semiconductor device results in virtually no loading of the frequency-determining RC network.

Figure 2-7 shows the circuit of a phase-shift af oscillator employing a single FET. In this circuit, the frequency is determined by the three-leg RC phase-shift network (C1-C2-C3-R1-R2-R3) from which the oscillator takes its name. For the necessary 180° phase shift for oscillation in the feedback path between the drain and gate of FET Q1, the R and C values in the network are selected for 60° shift in each leg (R1-C1, R2-C2, and R3-C3). For simplicity, the capacitances are kept equal (C1 = C2 = C3) and the resistances are kept equal (R1 = R2 = R3). The frequency of the network (and accordingly the oscillation frequency of the circuit) then is f = 1/(10.88 RC), where f is in hertz, R in ohms, and C in farads. With the network values given in Fig. 2-7, the frequency is 1021 Hz (for exactly 1000 Hz with the 0.05 μF capacitors, R1, R2, and R3 each must be 1838 ohms). When experimenting with a phase-shift oscillator, it will be easier to work with the resistors than with the capacitors. For an available capacitance (C), the required resistance (R) for a desired frequency (f) is R = 1/(10.88 f C), where R is in ohms, f in hertz, and C in farads. Thus, with the 0.05 μF capacitors shown in Fig. 2-7, the resistance required for 400 Hz = $1/(10.88 \times 400 \times 5 \times 10^{-8})$ = 1/0.0002176 = 4596 ohms.

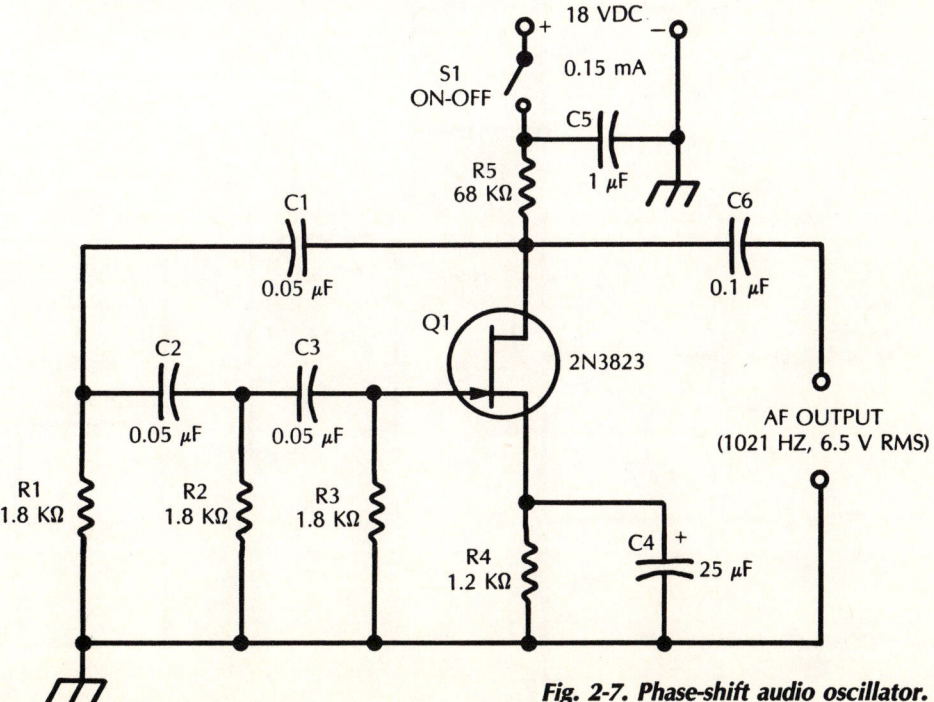

Fig. 2-7. Phase-shift audio oscillator.

The 2N3823 FET provides the high transconductance (6500 μmho) required for good operation of a phase-shift oscillator. The circuit draws approximately 0.15 mA from the 18-Vdc supply. The open-circuit af output is approximately 6.5 volts rms. (The dc and af values both will vary somewhat with individual FETs.)

All resistors in the circuit are ¼- or ½-watt. C5 and C6 can be any convenient low-voltage units. Electrolytic capacitor C4 is a 25-volt unit. For frequency stability, capacitors C1, C2, and C3 must be of top quality and closely matched in capacitance.

No. 6: Product Detector

Figure 2-8 shows the circuit of a simple product detector employing a HEP F0015 field-effect transistor. This detector is easily included in compact receiver circuits for CW and SSB reception. Here, the output of the i-f amplifier is applied to the gate of the FET, and the output of the beat-frequency oscillator (BFO) is applied to the source across unbypassed source resistor R4. The i-f and BFO signals are mixed and produce an af signal in the drain output circuit of the FET. A pi-type low-pass filter (C4-C5-RFC) removes the intermediate frequency from the audio output of the detector (a suitable 91-mH rf choke for use in this filter is J. W. Miller No. 70F912AF, having an internal resistance of 250 ohms).

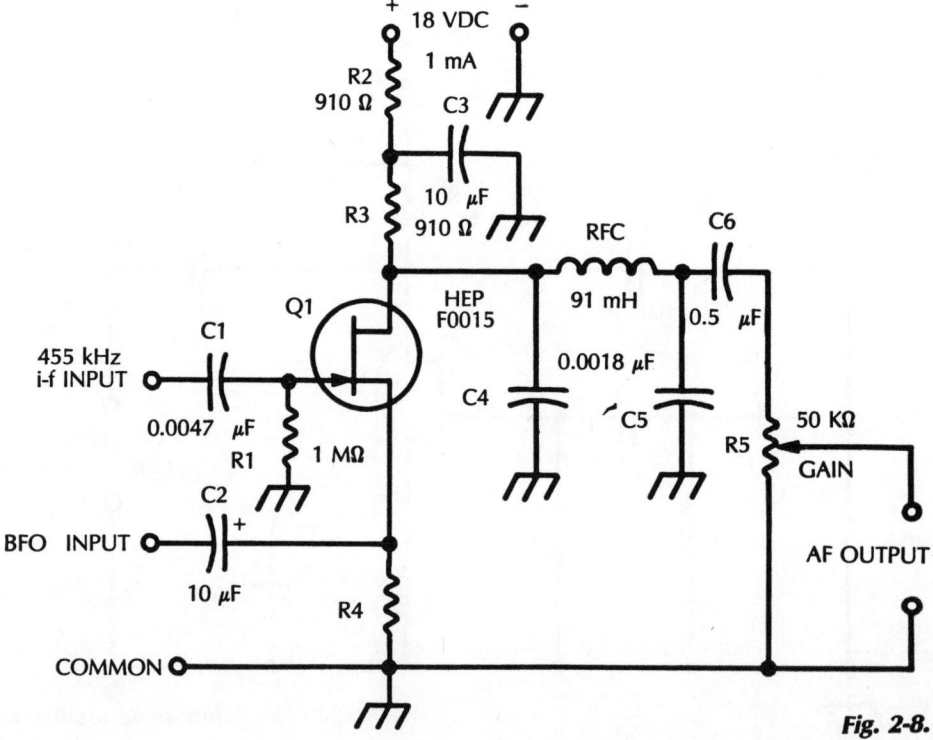

Fig. 2-8.

The circuit draws approximately 1 mA from the 18-Vdc supply which may be obtained from any convenient point in the receiver, but this current will vary somewhat with individual FETs. The af output amplitude will depend upon the relative amplitudes of the i-f and BFO signal voltages. For low-distortion operation, i.e., linear detection, it is recommended that the amplitude of the BFO signal be not less than five times that of the i-f signal. Ideally, there is no output unless the i-f and BFO signals are both applied. The resulting output signal is the product of the two input signals, hence the name product detector.

In construction, the i-f and BFO input sections of the detector should be kept as clear of each other as practicable, to minimize stray coupling; shielding is advisable. Electrolytic capacitors C2 and C3 are rated at 35 volts or higher; capacitors C1, C4, and C5 are mica or good-grade ceramic units; and capacitor C6 may be any convenient low-voltage unit. All resistors are ¼- or ½-watt.

No. 7: Regenerative Receiver

The regenerative receiver is a perennial favorite of radio experimenters. A reasonably sensitive device, it has a simple circuit and is easy to operate. It works over a wide frequency range and responds to either modulated or continuous-wave signals. For a long time it was the only receiver used by many radio hams and short-wave listeners. As an emergency receiver, the regenerative set can be quickly assembled at low cost.

Figure 2-9 shows the circuit of a tickler-coil type of regenerative receiver which covers the frequency range 440 kHz to 30 MHz in five overlapping bands: 440-1200 kHz, 1-3.5 MHz, 3.4-9 MHz, 8-20 MHz, and 18-30 MHz.

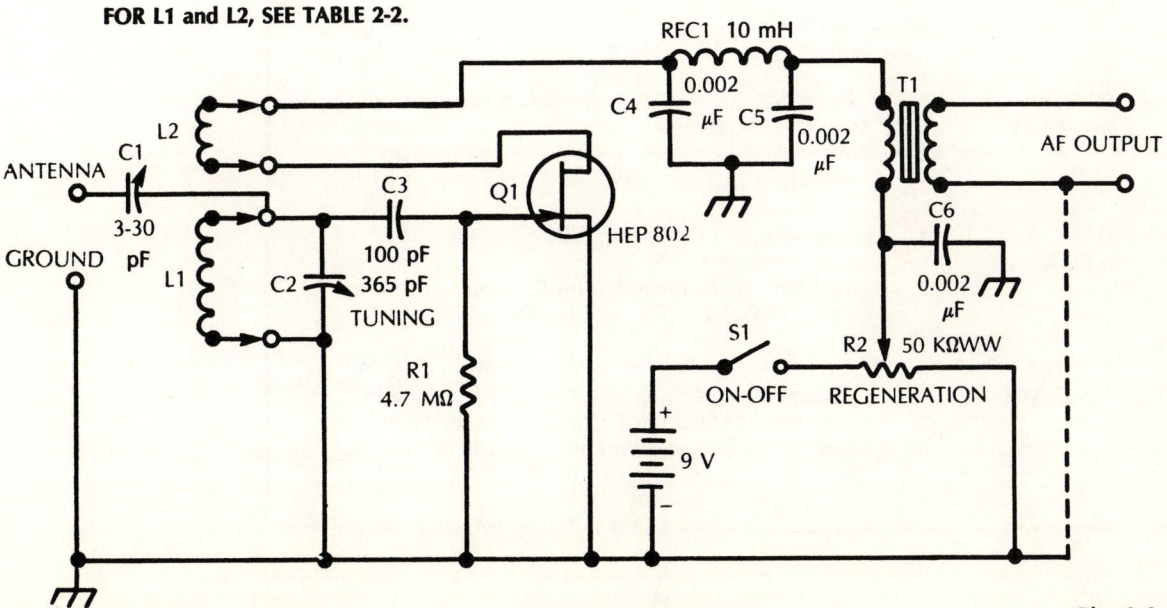

Fig. 2-9.

Plug-in coils provide a separate tuning inductor (L1) and tickler (L2) for each band. The single FET is an inexpensive HEP 802. In this circuit, positive (regenerative) feedback is obtained by inductively coupling energy back from the drain output circuit of the FET to the gate input circuit via the coupling between tickler coil L2 and tuning coil L1.

The regeneration control is the 50 kΩ wirewound potentiometer, R2, by means of which the dc drain voltage is adjusted. At the highest-voltage setting of R2, the circuit will break into oscillation. A pi-type filter (RFC1-C4-C5) removes the rf component from the output circuit. Resistor R1 is a ¼- or ½-watt unit, and capacitors C3, C4, C5, and C6 should be mica or good-grade ceramic units.

The audio-frequency output signal is coupled from the circuit by transformer T1. This may be any convenient interstage coupling transformer, preferably having a secondary-to-primary turns ratio of 1:1, 2:1, or 3:1. The receiver may be coupled to an audio amplifier, or high-impedance headphones may be connected directly to the af output terminals.

Table 2-1 gives specifications and winding instructions for the plug-in coils. These coils are wound on 1-inch diameter 4-pin plastic forms and are plugged into a 4-contact socket.

Band	L1	L2
BAND A 440-1200kHz	187 Turns No. 32 Enameled Wire Closewound on 1 In. Diameter Form.	45 Turns No. 32 Enameled Wire Closewound on Same Form as L1. Space 1/16" From Top of L1.
BAND B 1-3.5 MHz	65 Turns No. 32 Enameled Wire Closewound on 1 In. Diameter form.	15 Turns No. 32 Enameled Wire Closewound on Same form as L1. Space 1/16" From Top of L1.
BAND C 3.4-9 MHz	27 Turns No. 26 Enameled Wire Closewound on 1 In. Diameter Form.	8 Turns No. 26 Enameled Wire Closewound on Same Form as L1. Space 1/16" From Top of L1.
BAND D 8-20 MHz	10 Turns No. 22 Enameled Wire Closewound on 1 In. Diameter Form.	4 Turns No. 22 Enameled Wire Closewound on Same Form as L1. Space 1/16" From Top of L1.
BAND E 18-30 MHz	5½ Turns No. 22 Enameled Wire Airwound ½ In. in Diameter. Space to Winding Length on ½ In.	4 Turns No. 22 Enameled Wire Airwound ½ In. in Diameter. Space to Winding Length of ¼ In. Mount 1/16" From Top of L1.

Table 2-2. Coil-Winding Data.

To test the regenerative circuit initially:
1. Connect high-impedance headphones to af output terminals.
2. Set potentiometer R2 to its lowest-voltage position.
3. Set tuning capacitor C2 to maximum capacitance.
4. Close switch S1.
5. Slowly advance setting of R2; as maximum-voltage setting is approached, circuit should break into oscillation evidenced by beat-note whistle in headphones. If oscillation fails to occur, reverse leads to L1 or L2, but not both.
6. Connect antenna and ground, adjust C2 to tune-in signals, and adjust R2 for maximum weak-signal sensitivity without causing sustained oscillation of receiver. With some FETs, it may be necessary to experiment with the resistance of R1 to obtain maximum sensitivity.
7. The 3 to 30 pF antenna coupling capacitor, C1, is a screwdriver-adjusted trimmer capacitor. Set this trimmer for minimum capacitance that will give a strong signal. The adjustment should be optimum for all five frequency bands, so that C1 need not be readjusted.

The receiver tuning may be calibrated with the aid of an AM signal generator connected to the antenna and ground terminals. For this purpose, connect high-impedance headphones or af voltmeter to af output terminals. At each setting of the signal generator, tune C2 for audio peak and inscribe the frequency on the C2 dial. The top frequency in each band may be placed at the same point on the dial by setting the generator to that frequency and adjusting the trimmer on the frame of C2 (most 365 pF capacitors have this trimmer) for audio peak at that point. No volume control has been included in the circuit. If it is desired, a 50 kΩ potentiometer may easily be added at the af output terminals.

The current drawn by the receiver from the 9-Vdc source (B) depends upon the setting of potentiometer R2 and will be of the order of 1 mA at the highest setting (maximum regeneration).

No. 8: Superregenerative Receiver

For its small size, low cost, and simplicity, the superregenerative detector has no equal in sensitivity to signals. Especially useful at ultra-high frequencies, the superregenerator is broad enough in response to "hold onto" a floppy signal, and it has a built-in agc action.

Figure 2-10 shows the circuit of a self-quenching type of superregenerative receiver built around a 2N3823 VHF field-effect transistor. With four coils, the circuit covers the 2-, 6-, and 10-meter ham bands and the 27 MHz region (see Table 2-3). The frequency coverage allows the receiver to be used for general communications and for radio model control. All coils are single, two-terminal units. The 27 MHz and 6- and 10-meter coils are commercial, slug-tuned units which must be mounted on two-pin plugs for easy insertion and removal (for singleband receivers, these coils may be soldered permanently into the circuit).

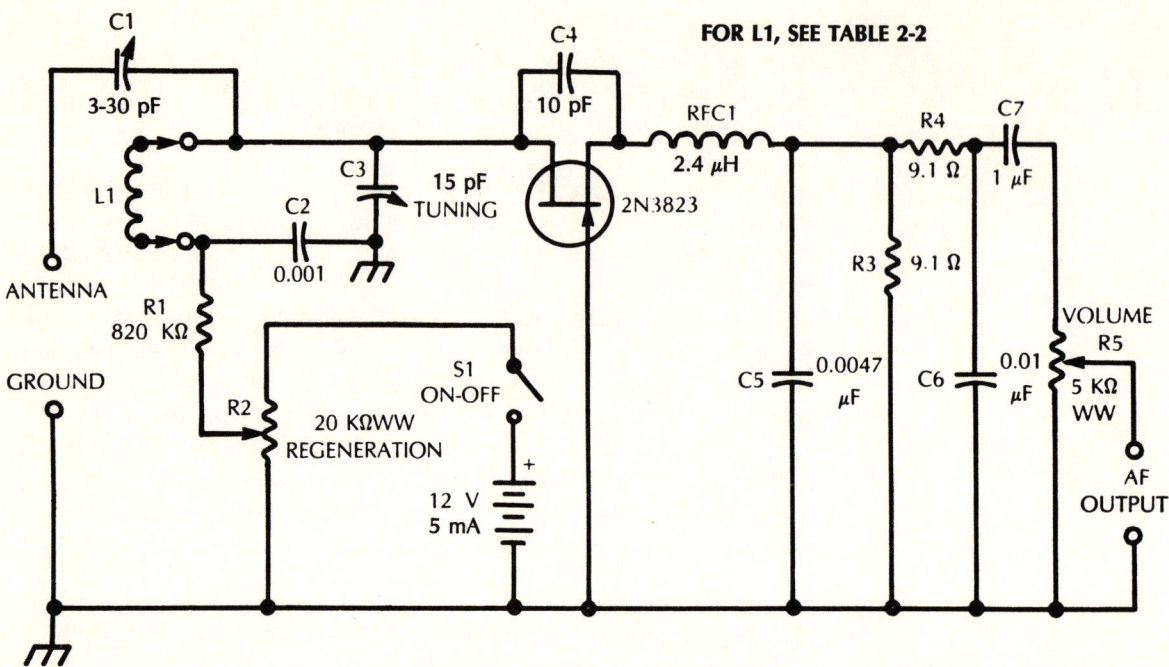

Fig. 2-10. Superregenerative receiver.

However, the 2-meter coil must be wound by the reader, and it too must be provided with a plug-in base, except in a single-band receiver. A filter system (RFC1-C5-R3) removes the rf component from the receiver output circuit, and a second filter (R4-C6) attenuates the quench frequency. A suitable 2.4 μH inductor for the rf filter is J. W. Miller No. 4406.

TUNING CAPACITANCE = 15 pF
- **A.** 10-Meter Amateur Band. J.W. Miller No. 4405. Adjust Core.
- **B.** 6-Meter Amateur Band. J.W. Miller No. 4403. Adjust Core.
- **C.** 2-Meter Amateur Band. 4 Turns No. 14 Bare Wire Airwound ½ In. in Diameter. Space to Hit Band.
- **D.** 27-MHz Range. J.W. Miller No. 4405. Adjust Core.

Table 2-3. Coil Data.

To test the superregenerative circuit initially:
1. Connect high-impedance headphones to af output terminals.
2. Set volume-control potentiometer R5 to its maximum-output position.
3. Set regeneration-control potentiometer R2 to its bottom end.
4. Set tuning capacitor C3 to maximum capacitance.
5. Close switch S1.

6. Advance setting of potentiometer R2, noting that at one setting a loud hiss begins, marking the onset of superregeneration. Loudness of this hiss should be fairly uniform as capacitor C3 is varied, but should increase somewhat as R2 is advanced to the top of its setting.
7. Connect antenna and ground. If the connection of an antenna stops the hiss, adjust the antenna trimmer capacitor C1 to restore this. This trimmer is screwdriver adjusted and should need adjustment only once to accommodate all frequency bands.
8. Tune-in signals in each band, noting the agc action of the receiver and the nasal quality of its voice reproduction.
9. The receiver tuning dial, attached to C3, may be calibrated with the aid of an AM signal generator connected to the antenna and ground terminals. Connect high-impedance headphones or af voltmeter to af output terminals, and at each setting of the generator, tune C3 for audio peak. The top frequency in the 10-meter, 6-meter, and 27 MHz bands may be placed at the same point on the C3 dial by adjusting the screw slugs in the corresponding coils, with the signal generator set to the appropriate frequency and with C3 set to the desired point near minimum capacitance. The 2-meter coil, however, has no slug and must be adjusted by squeezing or spreading its turns for alignment with the top-band frequency.

Because of the high frequencies at which this receiver operates, particular care is needed in its construction. The shortest practicable straight leads must be used in the tuned circuit (L1-C1-C2-C3-R1), and the mounting of this part of the receiver must be firm and solid. Any subpanel employed in the tuning section should be made of polystyrene. Capacitors C2, C4, C5, and C6 are mica units, but C7 may be any convenient low-voltage unit. Fixed resistors R1, R3, and R4 are ¼- or ½-watt composition or film units.

The user must remember that the superregenerative receiver is a notorious radiator of rf energy and can interfere with other nearby receivers tuned to the same frequency. The antenna coupling trimmer, C1, provides some attenuation of this radiation and so does reduction of the battery voltage to the lowest value that will still afford good sensitivity and audio volume. A radio-frequency amplifier operated ahead of the superregenerator is a very effective medium for minimizing radiation; but at the frequencies at which the receiver operates, such an amplifier is not easily fabricated and it negates the simplicity of the superregenerator.

No. 9: Electronic dc Voltmeter

Figure 2-11 shows the circuit of a balanced electronic dc voltmeter having an input resistance (including the 1-megohm resistor in the shielded probe) of 11 megohms. The circuit draws approximately 1.3 mA from a self-contained 9-volt battery. Therefore it can be left running for extended periods. This instrument covers 0-1000 volts in eight ranges: 0-0.5, 0-1, 0-5, 0-10, 0-50, 0-100, 0-500, and 0-1000 volts.

In the input voltage divider (range switching), the required resistances are made up of series-connected stock-value resistors which must be selected carefully for resistance close to the indicated values. If precision instrument-type resistors are available, the number of resistors in this string can be halved. That is, for R2 and R3, substitute 5 MΩ; for R4 and R5, 4MΩ; for R6 and R7, 500 kΩ; for R8 and R9, 400 k; for R10 and R11, 50 k; for R12 and R13, 40 k; for R14 and R15, 5 k; and for R16 and R17, 5 k.

A balanced circuit such as this has virtually no zero drift; any drift in FET Q1 is balanced out automatically by a similar drift in Q2. The internal drain-to-source circuits of the FETs, in conjunction with resistors R20, R21, and R22, form a resistance bridge. Indicating microammeter M1 is the detector in this bridge circuit. With zero signal input to the electronic voltmeter, meter M1 is set to zero by balancing this bridge with the aid of potentiometer R21. When a dc voltage is subsequently applied to the input terminals, the bridge unbalances—since the internal drain-to-source resistance of the FETs changes—and the meter deflects proportionately. The RC filter formed by R18 and C1 removes ac hum and noise picked up by the probe and the voltage-switching circuits.

Initial Calibration

With zero voltage at the input terminals:

1. Close switch S2 and set potentiometer R21 to zero pointer of meter M1. Range switch S1 may be set to any position for this step.
2. Set range switch to its 1-volt position.
3. Connect an accurately known 1-volt dc source to input terminals.
4. Adjust calibration control R19 for exact full-scale deflection of meter M1.
5. Temporarily remove input voltage and note if meter still is zeroed. If it is not, reset R21.
6. Work back and forth between steps 3, 4, and 5 until a 1-volt input deflects meter to full scale, and meter remains zeroed when input is removed.

Rheostat R19 will need no readjustment after this process unless its setting is disturbed or the instrument later needs recalibration. Zero-set potentiometer R21 will need only occasional resetting.

If range resistors R2 to R17 are accurate, this single-range calibration will be sufficient; all other ranges will automatically be in calibration. A special voltage scale can be drawn for the meter, or the existing 0-100 μA scale may be read in volts by mentally applying the multiplier on all but the 0-100 volt range.

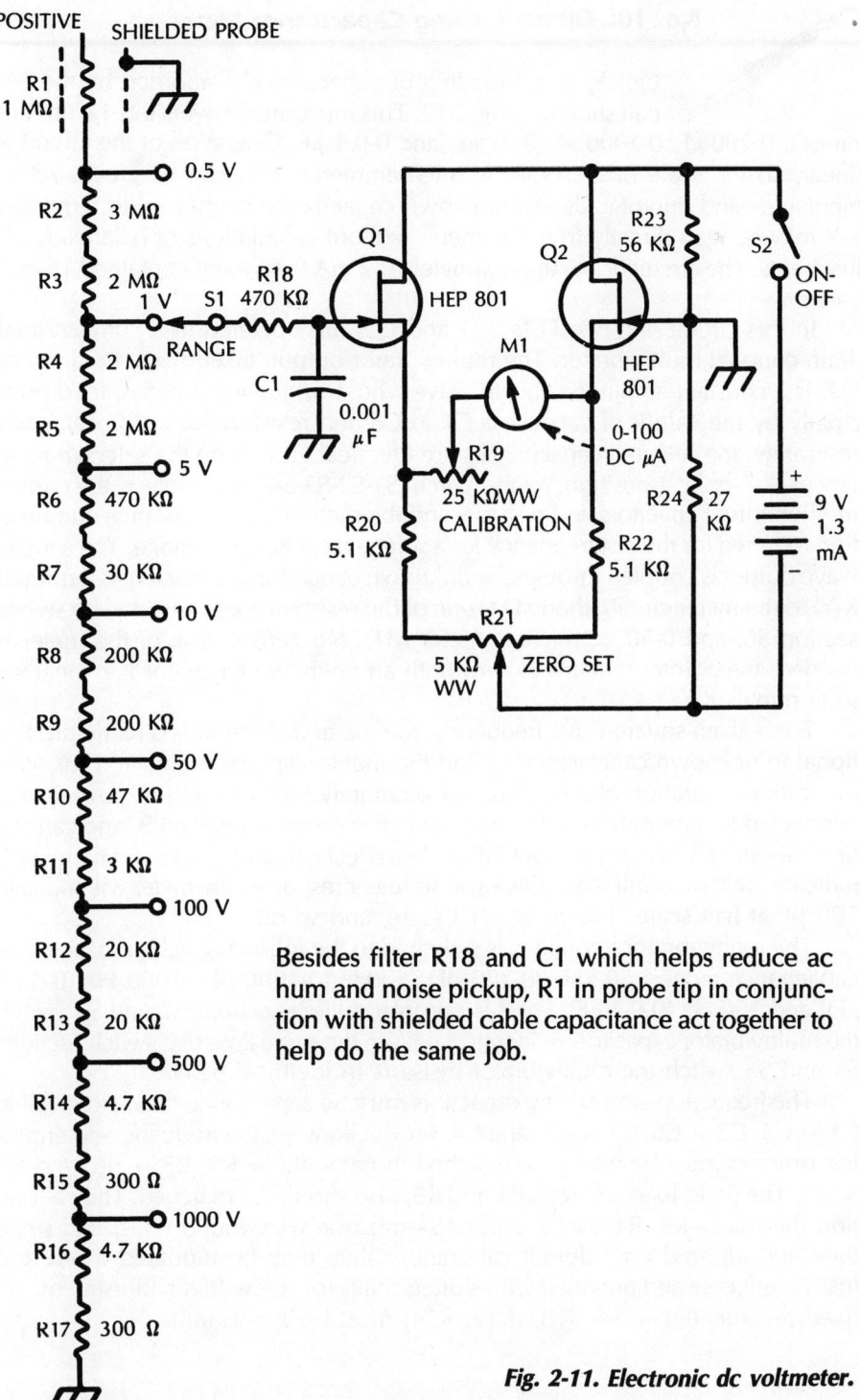

Besides filter R18 and C1 which helps reduce ac hum and noise pickup, R1 in probe tip in conjunction with shielded cable capacitance act together to help do the same job.

Fig. 2-11. Electronic dc voltmeter.

No. 10: Direct-Reading Capacitance Meter

Quick, direct reading of capacitance is afforded by the circuit shown in Fig.2-12. This instrument covers 0-0.1μF in four ranges: 0-200 μF, 0-1000 μF, 0-01 μF, and 0-0.1 μF. Operation of the circuit is linear, so the scale of the 0-50 dc microammeter. M1, can be graduated in picofarads and microfarads. An unknown capacitance connected to terminals X-Y may be read directly from the meter without calculations or balancing adjustments. The circuit draws approximately 0.2 mA from a self-contained 18-volt battery.

In this circuit, the two FETs (Q1 and Q2) are operated in a conventional drain-coupled multivibrator. The multivibrator output, taken from the drain of Q2, is a constant-amplitude square wave whose frequency is determined principally by the values of capacitors C1 to C8 and resistors R2 to R7. On each test range, the selected capacitances are identical and so are the selected resistances. A 6-pole, 4-position, rotary switch (S1-S2-S3-S4-S5-S6) selects the correct multivibrator capacitors and resistors and the meter-circuit resistance combination required for the test frequency for a selected capacitance range. The square-wave output is coupled through the unknown capacitor (connected to terminals X-Y) to the meter circuit (diode D1, one of the resistance legs selected by switch section S6, and 0-50 dc microammeter M1). No zero setting of the meter is needed; the pointer remains at zero until an unknown capacitor is connected to terminals X-Y.

For a given square-wave frequency, the meter deflection is directly proportional to unknown capacitance C, and the meter response is linear. Thus, if in the initial calibration of the circuit an accurately known 1000 pF capacitor is connected to terminals X-Y, the range switch is set to its position B, and calibration rheostat R11 is set for exact full-scale deflection of M1, then the meter will indicate 1000 pF at full scale. Owing to its linear response, the meter will indicate 500 pF at half scale, 100 pF at 1/10 scale, and so on.

The multivibrator frequency is switched to the following values for the four capacitance ranges: 50 kHz (0-200 pF), 5 kHz (0-1000 pF), 1000 Hz (0-0.01 μF), and 100 Hz (0-0.1 μF). For this purpose, switch sections S2 and S3 switch the multivibrator capacitors in identical pairs at the same time that switch sections S4 and S5 switch the multivibrator resistors in identical pairs.

The frequency-determining capacitors must be capacitance-matched in pairs: C1 = C5, C2 = C6, C3 = C7, and C4 = C8. Likewise, the frequency-determining resistors must be resistance-matched in pairs: R2 = R5, R3 = R6, and R4 = R7. The drain load resistors, R1 and R8, also should be matched. The calibration rheostats—R9, R11, R13, and R15—must be wirewound units; and since they are adjusted only during calibration, they may be mounted inside the instrument case and provided with slotted shafts for screwdriver adjustment. All fixed resistors (R1 to R8, R10, R12, R14) must be 1-watt units.

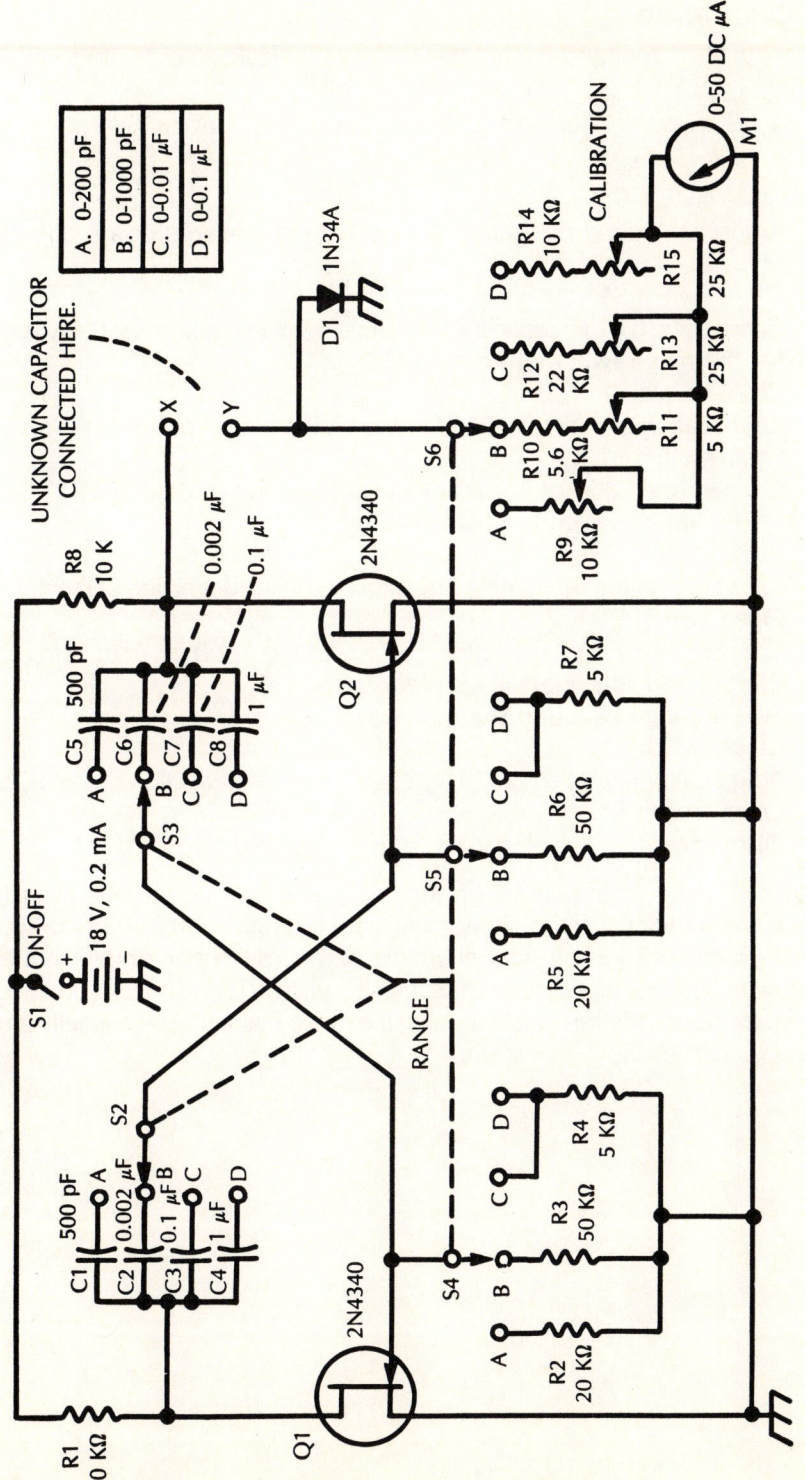

Fig. 2-12. *Direct-reading capacitance meter.*

Initial Calibration

For calibration, four accurately known, very-low-leakage capacitors will be required: 0.1 µF, 0.01 µF, 1000 pF, and 200 pF.

1. With the range switch in its position D, connect the 0.1 µF capacitor to terminals X-Y.
2. Close switch S1.
3. Adjust calibration rheostat R15 for exact full-scale deflection of meter M1.
4. Open switch S1.
5. Connect the 0.01 µF capacitor to terminals X-Y in place of the 0.1 µF unit.
6. Set the range switch to Position C.
7. Close switch S1.
8. Adjust calibration rheostat R13 for exact full-scale deflection of the meter.
9. Open switch S1.
10. Connect the 1000 pF capacitor in place of the 0.01 µF unit.
11. Set the range switch to position B.
12. Close switch S1.
13. Adjust calibration rheostat R11 for exact full-scale deflection of the meter.
14. Open switch S1.
15. Connect the 200 pF capacitor in place of the 1000 pF unit, using the shortest and straightest leads practicable.
16. Set the range switch to position A.
17. Close switch S1.
18. Adjust calibration rheostat R9 for exact full-scale deflection of the meter.
19. Open switch S1.
20. Disconnect the 200-pF capacitor from terminals X-Y.

A special meter scale may be drawn, or figures may be inscribed on the present microammeter scale to show capacitance ranges of 0-200 pF, 0-1000 pF, 0-0.01 µF, and 0-0.1 µF. In subsequent use of the instrument, simply connect an unknown capacitor to terminals X-Y, close switch S1, and read the capacitance from the meter. For best accuracy, use the range that will give the deflection in the upper part of the meter scale.

3
MOSFETs

The metal oxide semiconductor field-effect transistor (MOSFET) is a special type of field-effect transistor. The gate electrode is not the junction found in the conventional field-effect transistor but is instead a small metal plate insulated from the rest of the structure by a thin film of silicon dioxide. Any gate current in this device therefore is the leakage current of this insulation. It can be as low as 10 picoamperes (10^{-11} ampere) in some models. This gate gives the MOSFET an extremely high input resistance that is superior to that of the FET. This feature of the MOSFET is responsible for the alternate name of the device: insulated-gate field-effect transistor (IGFET).

MOSFETs are now available, as discrete units, in a wide variety of ratings and types.

THEORY

Figure 3-1A shows the cross section of a MOSFET. Although this representation may not be the exact picture of a particular manufactured unit, it is functionally correct and will serve to illustrate the main differences between MOSFETs and conventional junction FETs. (The various regions shown in the substrate are not to scale.) The electrodes of the MOSFET are drain, D (corresponding to the plate of a tube or the collector of a bipolar transistor), source, S (corresponding to the cathode of a tube or the emitter of a bipolar transistor), and gate, G (corresponding to the grid of a tube or the base of a bipolar transistor). The drain and source consist of small n-regions processed into a p-type substrate. Connecting the drain and source is a thin n-type channel lying just under the top face of the substrate. The metal gate electrode rests on the channel and is insulated from it by a very thin film of silicon dioxide (the insulating film is grown on the substrate, and the gate is deposited on the film).

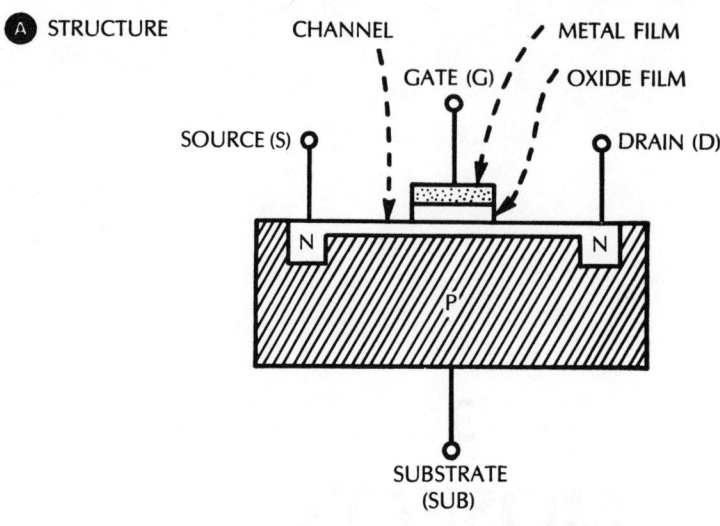

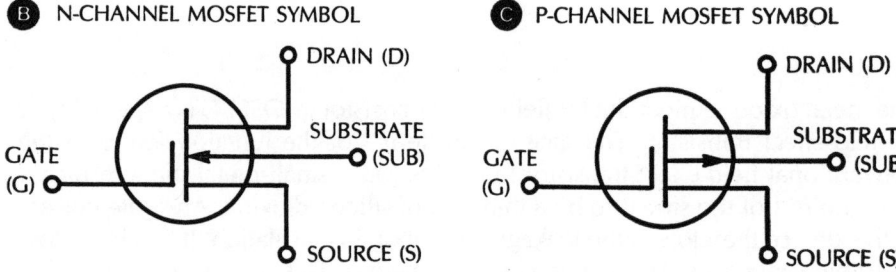

Fig. 3-1. Details of MOS field-effect transistor.

This particular type of device is termed an n-channel MOSFET, and its circuit symbol is shown in Fig. 3-1B. If, instead of this structure, we have an n-type substrate, the drain, source, and channel are p-type. The resulting device is termed a p-channel MOSFET, and its circuit symbol is shown in Fig. 3-1C. For the n-channel MOSFET, the external dc voltages are drain positive, source negative. For the p-channel MOSFET, the voltages are drain negative, source positive.

Figure 3-2 illustrates MOSFET operation. Here, a dc supply (V_{DD}) provides an operating voltage (V_{DS}) between drain and source, with drain positive and source negative; and a second dc supply (V_{GG}) provides a bias voltage (V_{GS}), with gate negative and source positive. The electric field of the gate penetrates the channel. Note that a gate-to-source resistor, R, is needed, since the gate insulation might be damaged if the gate floats.

In Fig. 3-2A, the gate voltage is zero; and under this condition, a maximum value of drain current (I_D) flows from supply V_{DD} through the channel, and back to V_{DD}, since there is no field from the gate to affect current flowing through the channel. In Fig. 3-2B, the gate voltage is negative; and the resulting field narrows the channel, reducing drain current I_D. When the gate voltage is raised to a sufficiently high negative value, the channel is depleted completely and the drain current is cut off.

Between these limits of maximum and cutoff, the drain current may be set to any intermediate value by appropriately setting the gate voltage. Figure 3-2C shows MOSFET performance in a family of curves resembling those for a pentode tube. An n-channel MOSFET is shown in Fig. 3-2; for a p-channel MOSFET, reverse the polarity of V_{DD} and V_{GG}.

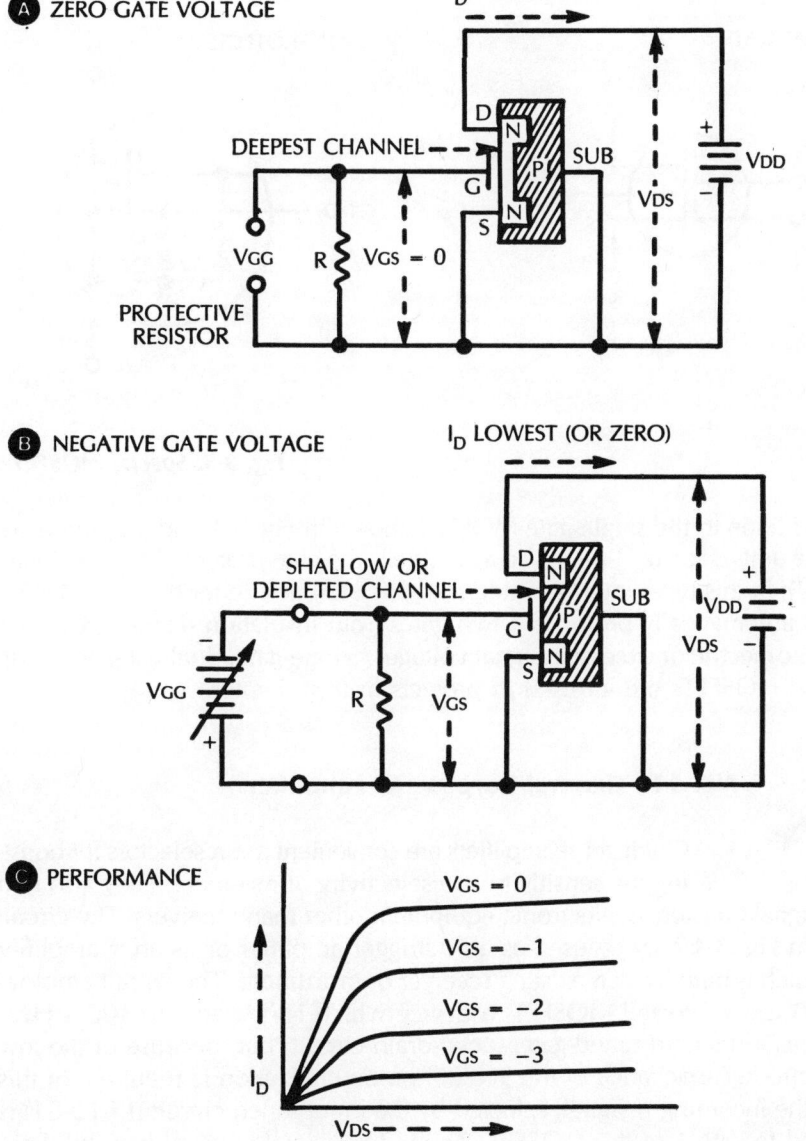

Fig. 3-2. MOSFET action.

Because the gate voltage acts to deplete the channel of current carriers, this type of MOSFET is termed depletion type. Other types are enhancement type and depletion/enhancement type, but they are not described here because all units mentioned in this chapter are of the depletion type.

Because the drain current of the MOSFET is controlled by gate voltage, the MOSFET possesses transconductance (up to 15,000 μmhos typical) and is a good amplifier. It also, like the tube and superior to the junction FET, has a very high input resistance (in some devices, up to 100 million megohms). Depending upon make and model, MOSFETs operate at frequencies up to 500 MHz.

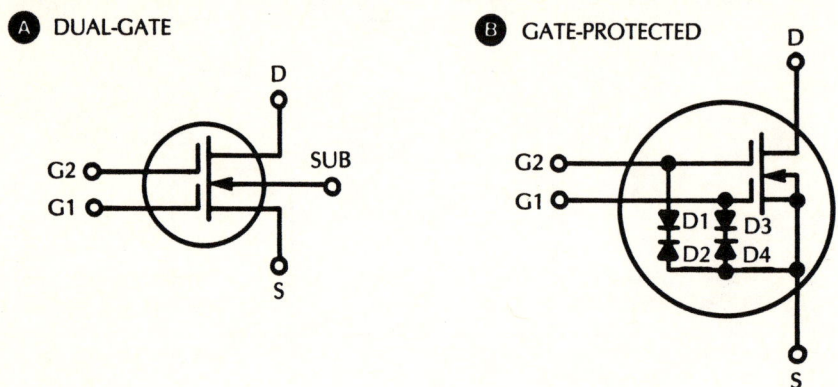

Fig. 3-3. Special MOSFETs.

In addition to the single-gate MOSFET shown in Fig. 3-1 and 3-2, there are dual-gate units (see Fig. 3-3A). Also available is the gate-protected MOSFET (see Fig. 3-3B); in this type, internal, integrated back-to-back zener diodes (D1, D2, D3, D4) automatically protect the two gates from insulation damage resulting from static electric or excessive signal voltages. Single-gate, dual-gate, and gate-protected MOSFETs are all used in projects in this chapter.

No. 11: General Purpose Rf Amplifier

Outrigger rf amplifiers are convenient as preselectors for boosting the sensitivity and selectivity of receivers. They also find use as signal boosters in electronic equipment other than receivers. The circuit shown in Fig. 3-4 may be used as an outrigger amplifier or as an rf amplifier stage which is built into an existing receiver or instrument. The circuit employs a 3N200 gate-protected MOSFET, a device which is operable to 400 MHz.

This is a standard tuned-gate, tuned-drain circuit; but, because of the low interelectrode capacitance of the 3N200, no neutralization is required. In this circuit, the incoming rf signal, selected by the input tuned circuit (L1-L2-C1) is presented to gate 1 of the MOSFET. The output signal is developed by drain current in the output tuned circuit (L3-L4-C5). Coaxial jacks (J1 and J2) or other

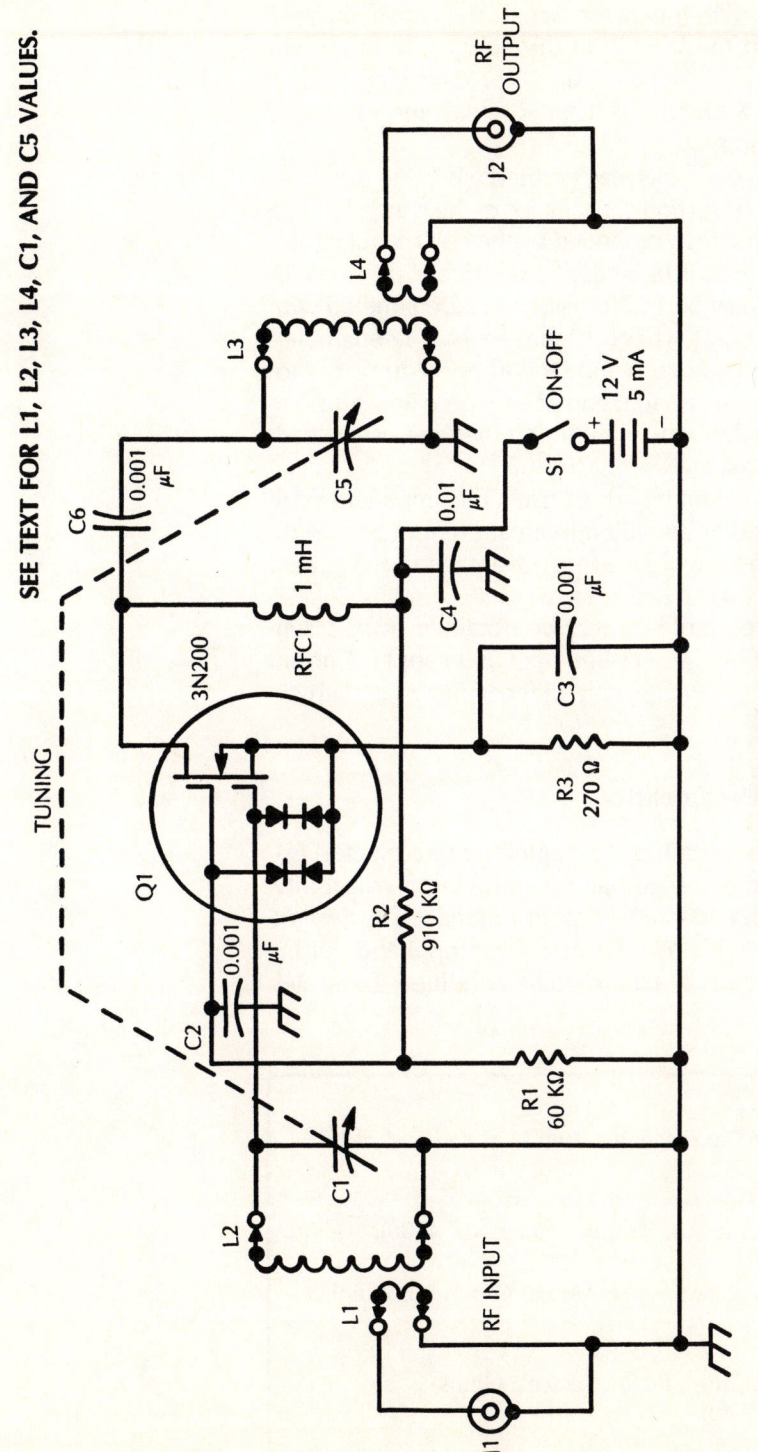

Fig. 3-4. General-purpose rf amplifier.

suitable terminals permit easy input and output of the signal. The negative gate-1 bias is provided by the voltage drop resulting from the flow of drain current through source resistor R3. Required positive gate-2 bias is produced by voltage divider R1-R2. With an individual MOSFET, resistor R2 may require some adjustment for maximum amplification.

Commercial plug-in coil sets with end links may be used when the amplifier must cover a wide tuning range. The per-section capacitance of the ganged tuning capacitor, C1-C5, then will be the value recommended by the coil manufacturer for the selected coils. When the amplifier is to be used exclusively in the standard broadcast band, L2 and L3 each must be 130 turns of No. 32 enameled wire closewound on a 1-inch diameter form. L1 will be 10 turns of No. 32 enameled wire closewound around the bottom end of L2, and L4 will be 12 turns of No. 32 enameled wire closewound around the bottom end of L3. These link-coupling coils (L1 and L4) are insulated from the coils with which they are associated. For the broadcast band, C1 and C5 each will be 365 pF.

At frequencies much above the standard broadcast band, the amplifier circuit should be shielded. The unit may readily be built into an aluminum box. With shielded construction of this kind, there will be some advantage in using feed through capacitors for C2, C3, and C4.

The dc operating voltage for the amplifier may be obtained from a self-contained battery or from a well-filtered, power-line-operated supply. Current drain is approximately 5 mA, but this value can differ somewhat with an individual 3N200.

No. 12: Ten-Meter Preamplifier

Figure 3-5 shows the circuit of a preamplifier (preselector) expressly for the 10-meter ham band. Employing permeability input and output tuning, this circuit is adapted from an original RCA design. Operation of the circuit is reasonably wide band, especially if input and output are stagger tuned, so that it can function as an aperiodic amplifier. Table 3-1 gives coil data.

L1   2 Turns No. 20 Enameled Wire Closewound Around Ground End of L2.

L2 AND L3   Each Subminiature, Slug-Tuned, Ceramic- Form Coil. Inductance Range 1.6 to 2.4 μH (J.W. Miller No. 4306).

L4   2 Turns No. 20 Enameled Wire Closewound Around Ground (Bottom) End of L3.

RFC1   6.28 μH Microminiature rf Choke, dc Resistance 2 Ohms (J.W. Miller No. 9230-40).

Table 3-1. Coil Data for 10-Meter Preamplifier.

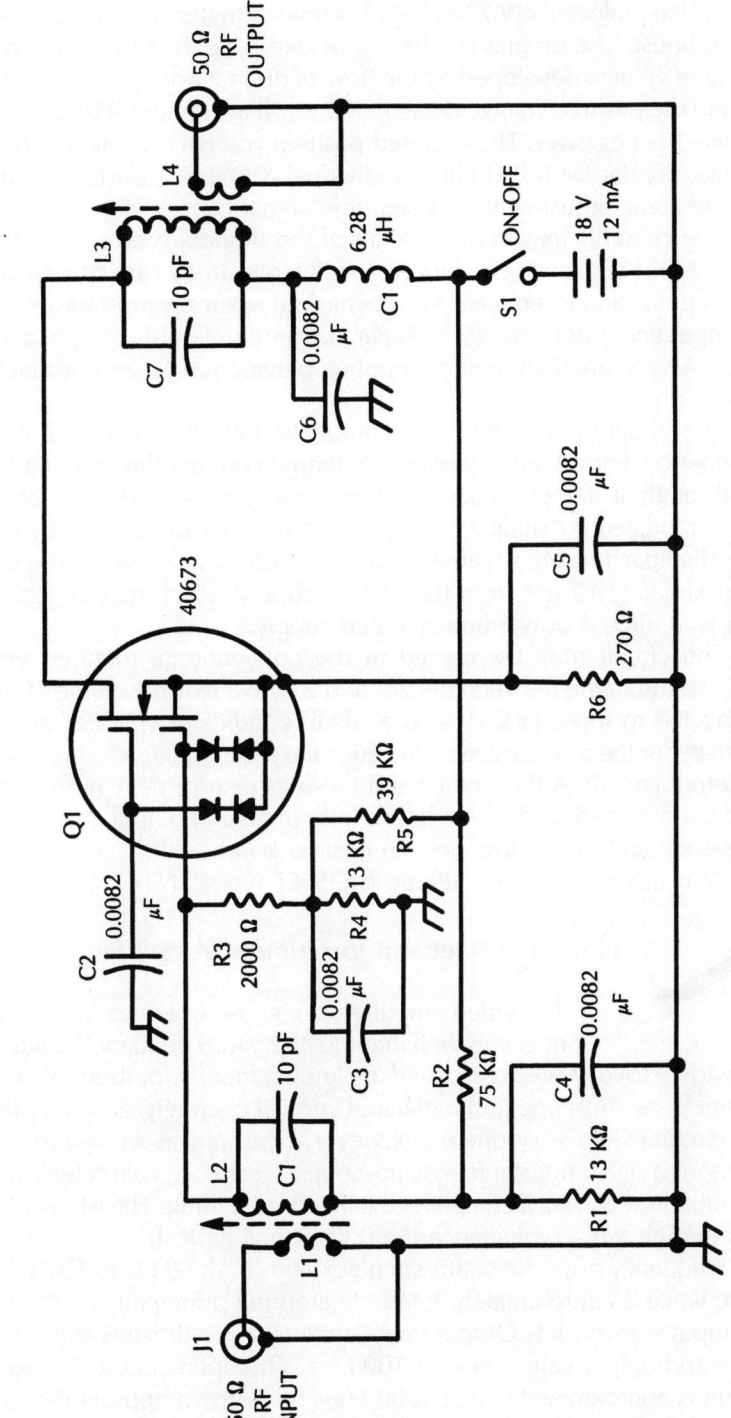

Fig. 3-5. Ten-meter preamplifier.

A gate-protected 40673 MOSFET is used with dual bias to provide substantial signal boost. The dc bias on the signal gate (gate 1) is the combination of the negative voltage developed by the flow of drain current through source resistor R6 and the positive voltage developed by voltage divider R1-R2 (the net voltage at gate 1 is negative). The required positive voltage on gate 2 is developed by voltage divider R4-R5. With an individual MOSFET, resistors R2 and R5 may require some adjustment for maximum signal output.

The circuit is fix-tuned by means of the slug-adjusted coils, L2 and L3 (see Table 3-1) While, owing to the very low interelectrode capacitance of the 40673, no neutralization is required, the mechanical assembly must be arranged to keep the input and output coils well separated and preferably at right angles to each other. A very small amount of coupling between input and output will induce oscillation.

The input and output accommodate low antenna and receiver-input impedance. For a high-impedance antenna, connect the antenna to the top of L2 through a 10 pF capacitor. Impedances other than 50 ohms may be accommodated by suitably changing the number of turns in L1 or L2.

All capacitors are silvered mica. All resistors are ½-watt. The circuit draws approximately 12 mA from the 12-Vdc supply which may be either a battery or a well-filtered power-line-operated supply.

The circuit may be aligned in the conventional manner, with a signal generator (tunable between the 28- and 29.7-MHz limits of the 10-meter band) connected to input jack J1, and a suitable indicator (such as an electronic rf voltmeter or the receiver itself into which the preamplifier is to operate) connected to output jack J2. Adjust the L2 and L3 slugs first for peak output at 28.8 MHz (approximate midband); then detune both input and output coils, one above this frequency and one below, for the desired bandwidth.

Also usable in this circuit are MOSFET types 3N187 and 3N200.

No. 13: Wideband Instrument Amplifier

The video amplifier in Fig. 3-6 operates from low audio frequencies well into the high radio frequencies. Such amplifiers are widely used in electronics and are known mainly for their role in TV picture channels, oscilloscope horizontal and vertical channels, and sampling systems. The circuit in Fig. 3-6 is offered, however, especially as an instrument amplifier; that is, as a signal booster for oscilloscopes, electronic voltmeters, and other test instruments requiring an amplifier that needs no tuning. The MOSFET is a 40820, having high transconductance (12,000 µmhos typical).

Frequency response of this circuit extends from 60 Hz to 10 MHz. The input impedance is approximately 1 MΩ, determined principally by the resistance of the input resistor, R1. Output impedance is approximately 2800 ohms. These input and output values hold at 1000 Hz. The open-circuit voltage gain of the circuit is approximately 10 at 1000 Hz, and is down approximately 3 dB at 60 Hz and down approximately 6 dB at 10 MHz. The maximum input-signal voltage before output-signal peak clipping is approximately 100 mV rms, and the corresponding maximum output-signal voltage is approximately 1 volt rms.

The peaking elements are trimmer capacitor C3 (55-300 pF) and slug-tuned inductor L1 (24-35 μH, J.W. Miller No. 4508 or equivalent). Both of these components are adjusted carefully for maximum gain of the amplifier at 10 MHz.

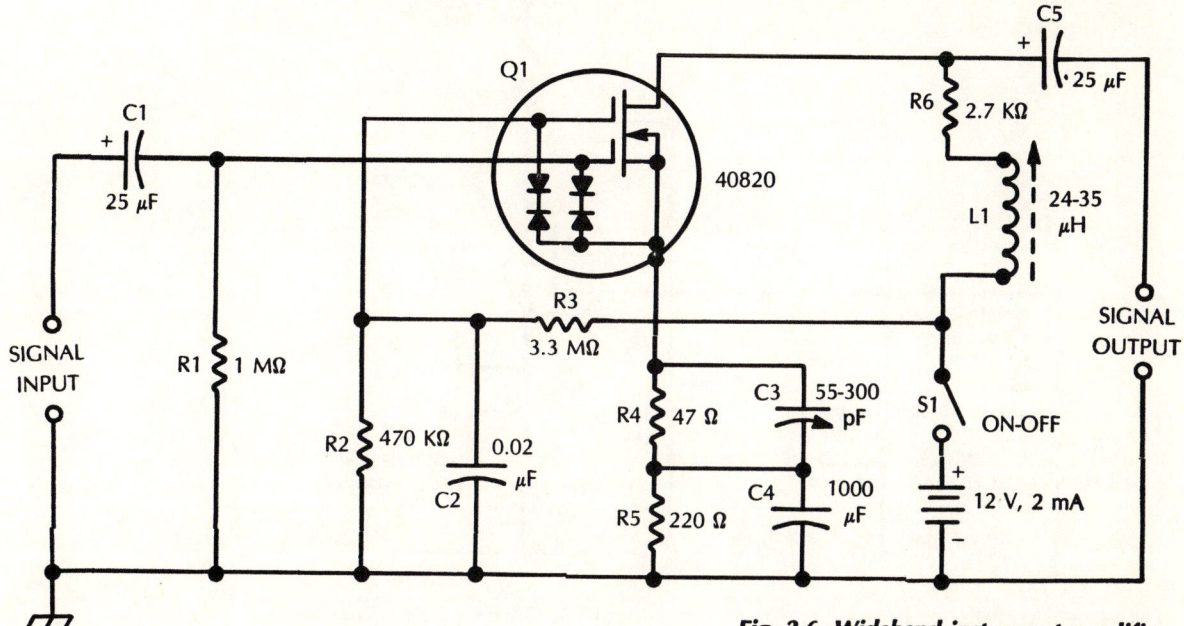

Fig. 3-6. Wideband instrument amplifier.

Successful high-frequency performance of the circuit demands that all wiring be short, rigid, and direct, and that leads be dressed for optimum performance and minimum cross coupling.

In this circuit, the MOSFET receives its negative gate-1 bias from the voltage drop resulting from the flow of drain current through source resistors R4 and R5 in series, and its positive gate-2 bias from voltage divider R2-R3, rf-bypassed by capacitor C2. With an individual MOSFET, resistance R3 may require some adjustment for maximum gain at 1000 Hz. Current drain from the 12-volt supply is approximately 2 mA, but this may vary somewhat with individual MOSFETs.

No. 14: Source Follower

Figure 3-7 shows the circuit of a compact source follower employing a HEP F2004 dual-gate MOSFET. This circuit is equivalent to the vacuum-tube cathode follower and the bipolar-transistor emitter follower. Like the latter two circuits, the source follower is a degenerative amplifier employing current feedback to cancel stage-introduced distortion. The circuit is characterized by high input impedance (in the MOSFET, this is higher than in either the bipolar or the JFET) and low output impedance. As such, it has many well-known applications in electronics, especially that of an impedance transformer with power gain. The source follower, like the cathode follower and emitter follower, is noted also for its wide frequency response and low distortion.

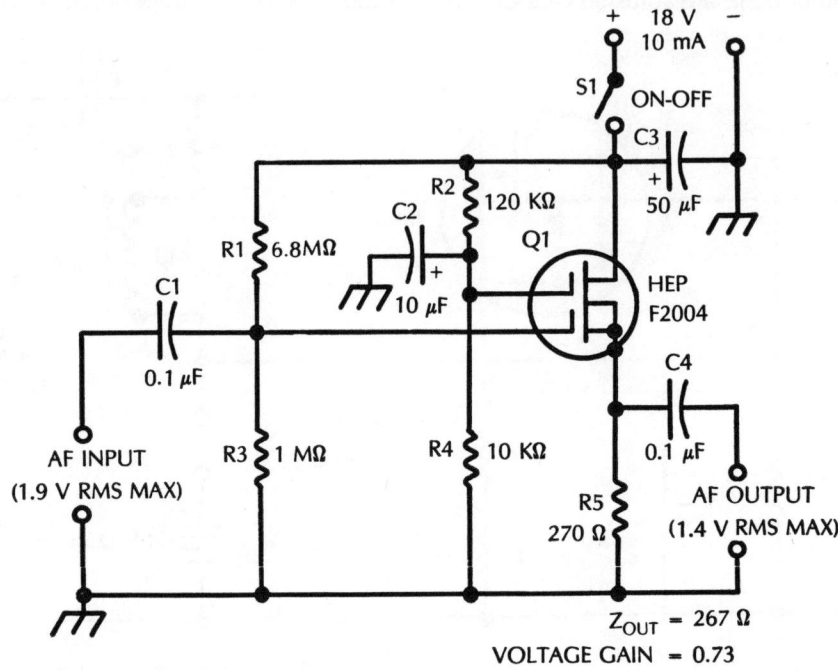

Fig. 3-7. Source follower.

In this circuit, the negative dc bias on gate 1 is the combination of the negative voltage developed by the flow of drain current through source resistor R5 and the positive voltage developed by voltage divider R1-R3 (the net voltage at gate 1 is negative). The required positive bias voltage on gate 2 is developed by voltage divider R2-R4. With an individual MOSFET, resistances R1 and R2 may require some adjustment for maximum signal output.

The input resistance of the circuit is approximately 1 megohm, and is largely determined by input resistor R1. Higher resistance may be used, with some risk of stray pickup. The effective output impedance of the stage is 267.2 ohms; this will vary somewhat with individual MOSFETs, since transconductance is a factor in the determination of output impedance, and this varies in MOSFETs of the same type.

Voltage gain of the circuit is approximately 0.73. The maximum input-signal voltage before output-signal peak clipping is 1.9-volts rms, and the corresponding maximum output-signal voltage is 1.4-volts rms. The circuit draws approximately 10 mA from the 18-Vdc source, but this may vary somewhat with individual MOSFETs.

All resistors are ½-watt. Electrolytic capacitors C2 and C3 are 25-volt units; capacitors C1 and C4 may be any convenient low-voltage units.

No. 15: Q Multiplier

A Q multiplier is a special oscillator that is connected across a tuned circuit in a receiver to increase the latter's selectivity. The Q multiplier thus effectively increases the Q of the tuned circuit. The oscillator may be added either to an rf stage or i-f stage. Figure 3-8 shows the circuit of a Q multiplier which can be used in conjunction with a 455 kHz i-f stage.

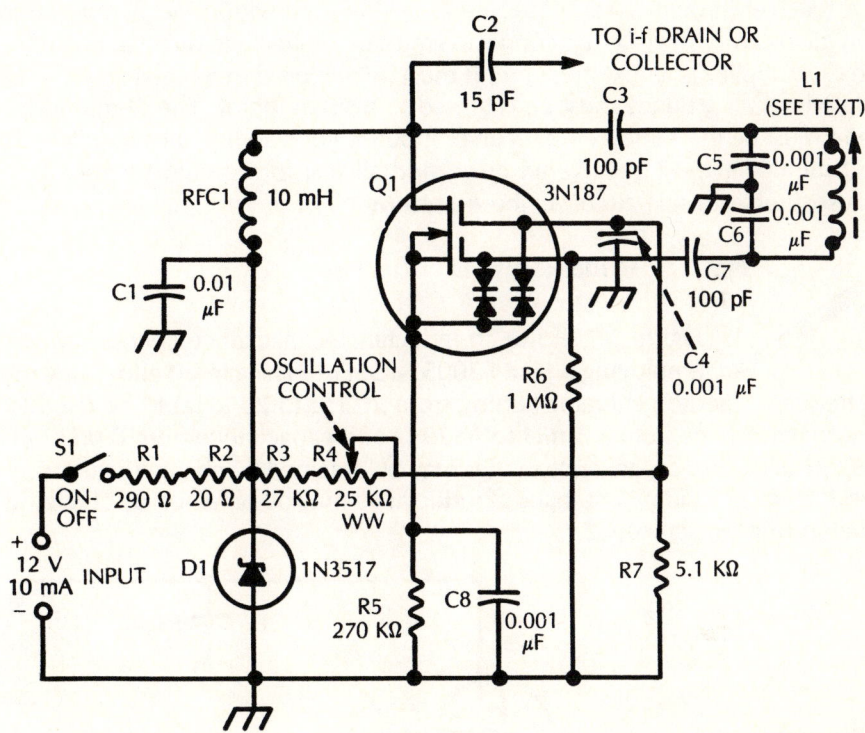

Fig. 3-8. (455 kHz i-f).

This circuit is basically a Colpitts i-f oscillator employing a 3N187 gate-protected MOSFET; it is connected to the i-f stage in the receiver through capacitor C2. The Colpitts oscillator employs a single inductor (L1) and split tuning capacitor (C5 and C6) in series. With the resulting 0.005 µF capacitance, the inductance of L1 (J.W. Miller No. 21A224RBI or equivalent) can be set, by means of the tuning slug, to 244 µH for oscillation at 455 kHz. The strength of oscillation is governed by the setting of the 25 kΩ wirewound oscillation control potentiometer, R4. At one setting of this control, the signal in the receiver is caused to peak sharply; at another setting of R4, a slot is created, and the signal is eliminated.

In this circuit, gate 1 of the MOSFET receives its negative bias as the voltage developed by the flow of drain current through source resistor R5. Gate 2 receives its positive bias from voltage divider R3-R4-R7. The dc voltage presented to the circuit is regulated by a 1N3517 zener diode in conjunction with resistors R1 and R2 at the constant value of 9.1 volts. (See Chapter 8 for treatment of zener diodes.) When fully oscillating, the circuit draws approximately 10 mA from the 12-Vdc supply.

Rf output from the circuit is coupled to the drain or collector of the i-f stage in the receiver through 15 pF capacitor C2. All wiring within the Q multiplier and in the receiver to which it is connected must be as short and direct as possible. If the Q multiplier is an external unit, it must be enclosed in a metal shield box inside which all grounds must be returned to a single point. The Q-multiplier ground must be the same as the receiver ground. For stability, all capacitors in the Q multiplier must be silvered mica, and all resistors should be 1-watt.

Also usable in this circuit is the 40820 MOSFET.

No. 16: Wide-Range LC Checker

Figure 3-9 shows an inductance-capacitance checker circuit employing a HEP F2005 MOSFET. This circuit allows a variable-frequency audio generator (tuning from 20 Hz to 20 kHz) to be used to measure inductance from 6.3 mH to 6329 H and capacitance from 0.0632 µF to 63,291 µF. Other capacitance ranges may be obtained by changing the value of inductance L1, and other ranges of inductance may be obtained by changing the value of capacitance C2.

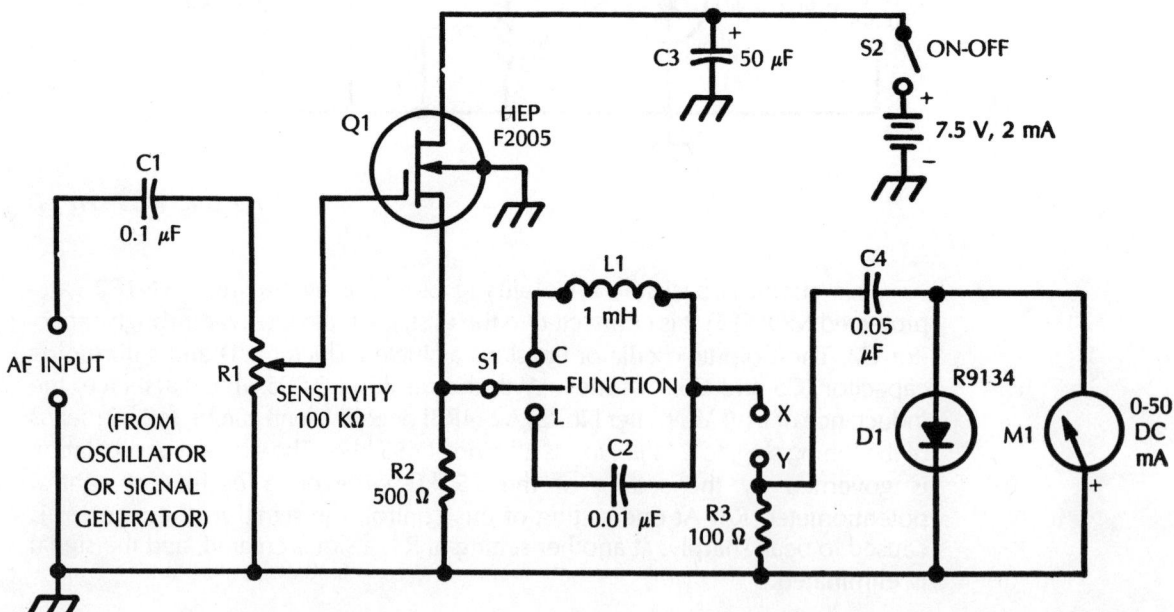

Fig. 3-9.

The MOSFET is connected in a source follower circuit which drives a series-resonant circuit containing the unknown capacitance connected in series with a known inductance, L1, or the unknown inductance connected in series with a known capacitance, C2. The generator is tuned for peak deflection of meter M1, which is part of an ac milliammeter circuit connected in the series-resonant circuit. At that point, the generator frequency is noted, and the unknown component calculated in terms of the known values. Thus,

$$C_x = 1/(39.5 f^2\ L1)$$

and

$$L_x = 1/(39.5 f^2\ C2).$$

The unknown component is connected to terminals X-Y. Resonant current flowing through R3 develops a voltage drop across this resistor, and this voltage is rectified by diode D1 to deflect meter M1. The standard inductor (L1) is a 1 mH, slug-tuned coil (J. W. Miller No. 22A103RBI or equivalent) which may be set exactly to 1 mH with the aid of an inductance bridge. Standard capacitor C2 is a 0.01 μF silvered-mica unit which must be obtained with the highest accuracy possible.

Capacitance Measurement

To use the instrument for determining capacitance values, follow this procedure:

1. Connect the signal generator to the af input terminals.
2. Connect the unknown capacitor to terminals X-Y.
3. Set switch S1 to C.
4. Close switch S2 and turn on generator.
5. Tune generator throughout its range, watching for a peak deflection of meter M1.
6. At peak, read frequency from generator dial.
7. Calculate the capacitance: $C = 1/(0.0395\ f^2)$, where C is in farads and f in hertz. Note that the formula is simplified, since the inductance of L1 is constant.

Inductance Measurement

To use the instrument for determining inductance values, follow this procedure:

1. Connect the signal general to the af input terminals.
2. Connect the unknown inductor to terminals X-Y.
3. Set switch S1 to L.
4. Close switch S2 and turn on generator.
5. Tune generator throughout its range, watching for a peak deflection of meter M1.
6. At peak, read frequency f from generator dial.

7. Calculate the inductance: $L = 1/(.395 f^2 \times 10^{-6})$, where L is in henrys and f in hertz. Note that the formula is simplified, since the capacitance of C2 is constant.

In this circuit, the MOSFET receives its negative gate bias from the voltage drop resulting from the flow of drain current through source resistor R2. The sensitivity control potentiometer, R1, is set for best deflection of meter M1. Electrolytic capacitor C3 is a 25-volt unit, C2 is silvered mica, and C1 and C4 may be any convenient low-voltage units. For best stability, resistors R2 and R3 should be 1-watt.

The circuit draws approximately 2 mA from the 7.5-volt dc supply, but this current may vary somewhat with individual MOSFETs.

No. 17: Interval Timer

Figure 3-10 shows the circuit of a conventional interval timer in which the very high input resistance of the 3N142 MOSFET affords performance comparable to that of the usual vacuum-tube timer. Because of the smallness of the MOSFET, the relay, the 1000 µF (3V) capacitor, and other components, the entire unit can be very small and compact.

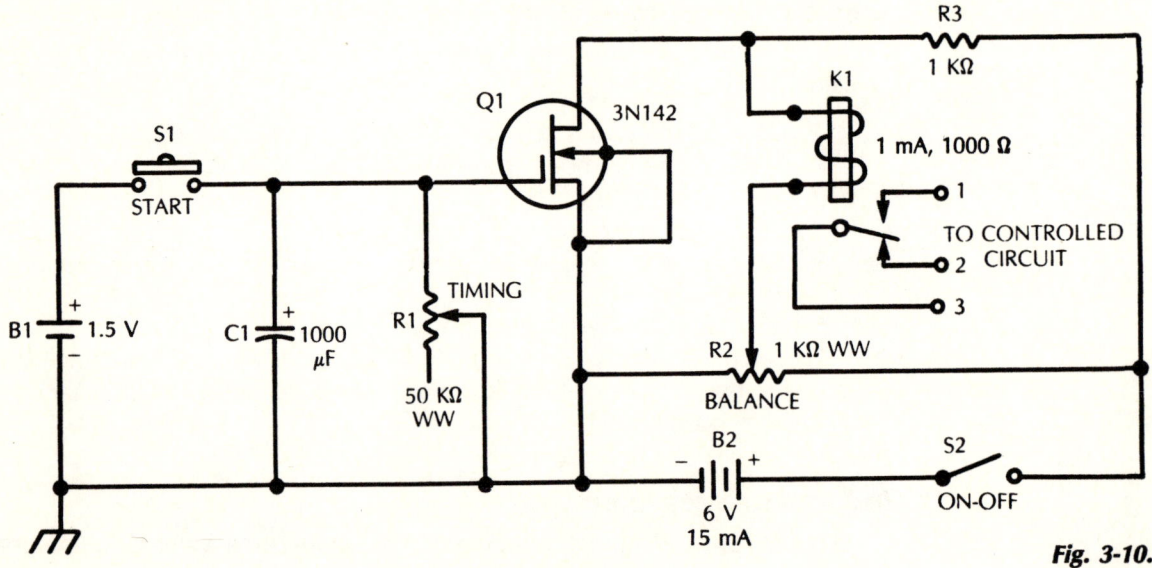

Fig. 3-10.

In this arrangement, the single-gate, 3N142 MOSFET serves as a simple dc amplifier driving relay K1 in its drain output circuit. When pushbutton switch S1 is depressed momentarily, capacitor C1 is charged from the 1.5-volt cell. At the same time, relay K1 closes. When S1 then is released, capacitor C1 discharges at a rate determined by capacitance C1 and the resistance setting of R1 (the time interval before the relay drops out is closely equal to the time constant of the circuit where t = R1C1 seconds).

The time instant at which the relay drops out may be controlled by proper setting of rheostat R1. With the 50,000-ohm rheostat and 1000 µF capacitor shown, the maximum time interval is 50 seconds. Other maximum values may be obtained by appropriate selection of C1 and R1 values in accordance with the time-constant formula.

The relay is a 1 mA, 1000-ohm device (Sigma 5F-1000 or equivalent) operated in a four-arm bridge in the drain circuit of the MOSFET. The bridge—consisting of resistor R3, the two "halves" of balance control potentiometer R2, and the internal drain-to-source resistance of the MOSFET—allows the static drain current to be balanced out of the relay (by adjustment of potentiometer R2) before the timer is ready to operate.

When the relay is closed, current drawn from the 6-Vdc supply is approximately 15 mA, but this may vary somewhat with individual MOSFETs. Controls R1 and R2 both are wirewound. Resistor R3 is ½-watt. The 1.5-volt cell, B1, is used only intermittently, to supply charging current to capacitor C1, and should enjoy long life, especially if a size-D cell is used.

No. 18: Capacitance Relay

Capacitance relays are found in a number of electronic systems. They are found also in nonelectronic areas, where they serve as object- or people-counters, intrusion alarms, proximity detectors, safety switches, crowd stoppers in exhibitions and displays, and so on. A solid-state capacitance relay is both small and fast-acting.

Figure 3-11 shows the circuit of a capacitance relay employing a 3N128 MOSFET in the sensitive rf oscillator stage and a 2N2716 silicon bipolar transistor in the high-gain dc relay-amplifier stage. In this arrangement, MOSFET Q1 operates as an oscillator in conjunction with inductor L1, trimmer capacitor C2, and radio-frequency choke RFC1 (the inductor, trimmer, and rf choke are self-contained in a commercial capacitance relay coil, J. W. Miller No. 695 or equivalent). The MOSFET stage oscillates lightly; and when a hand or other part of the human body comes near the pickup antenna (a short piece of wire with or without a metal plate on its end), the drain current of the MOSFET undergoes a slight change. This change is coupled to the input circuit of the 2N2716 dc amplifier which, in turn, closes the relay. When the hand is withdrawn, the events reverse and the relay opens.

To set up the device initially, protect the pickup antenna from any nearby bodies, including the operator's. Then, work back and forth between adjustments of threshold control R3 and sensitivity control R2 until the relay just opens. Next, bring one of your hands close to the antenna and alternately adjust trimmer C2 and rheostat R2 with the other hand until the relay just closes. Withdrawing your hand from the vicinity of the antenna should then cause the relay to open. When the circuit is correctly adjusted, bringing the hand to a desired distance from the antenna should close the relay, and moving the hand away should reopen it. The circuit must not be so sensitively adjusted that it self-operates.

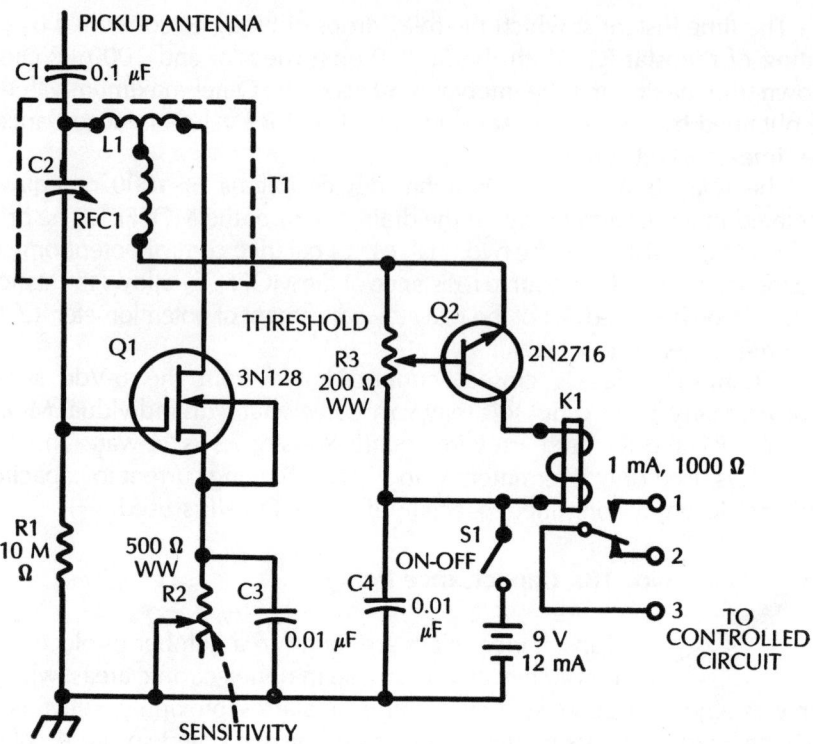

Fig. 3-11. Capacitance relay.

For reliable operation, the device should be built solidly and should be completely enclosed in a shield box, except for the pickup antenna. The assembly should be grounded to earth. The antenna must be steady. Capacitor C3 should be a mica or good-grade ceramic unit, but capacitors C1 and C4 can be any convenient low-voltage units. Controls R2 and R3 both are wirewound.

The relay is a 1 mA, 1000-ohm device (Sigma 5F-1000 or equivalent). When the relay is closed, current drain from the 9-volt dc supply, B1, is approximately 12 mA, but this may vary somewhat with individual 3N128s and 2N2716s. Use output terminals 1 and 3 when relay closure must close an external circuit; use output terminals 2 and 3 when relay closure must open an external circuit.

No. 19: Touch-Plate Relay

The high input resistance, high transconductance, and low input capacitance of the MOSFET suit this device for use as a dc amplifier in touch-plate relay circuits. Figure 3-12 shows a simple touch-plate circuit that responds to a light touch of a finger. Such relays have become common for operating all sorts of electrical gear, from sophisticated electronic circuits to elevators. A 3N152 low-noise MOSFET is employed in the circuit shown here. The gate almost floats, being 22 megohms above ground through resistor

R1, and is connected to a small, metal, pickup disc. The gate thus is highly susceptible to pickup, so that merely touching it with a finger tip couples in enough random energy from stray fields to drive the MOSFET.

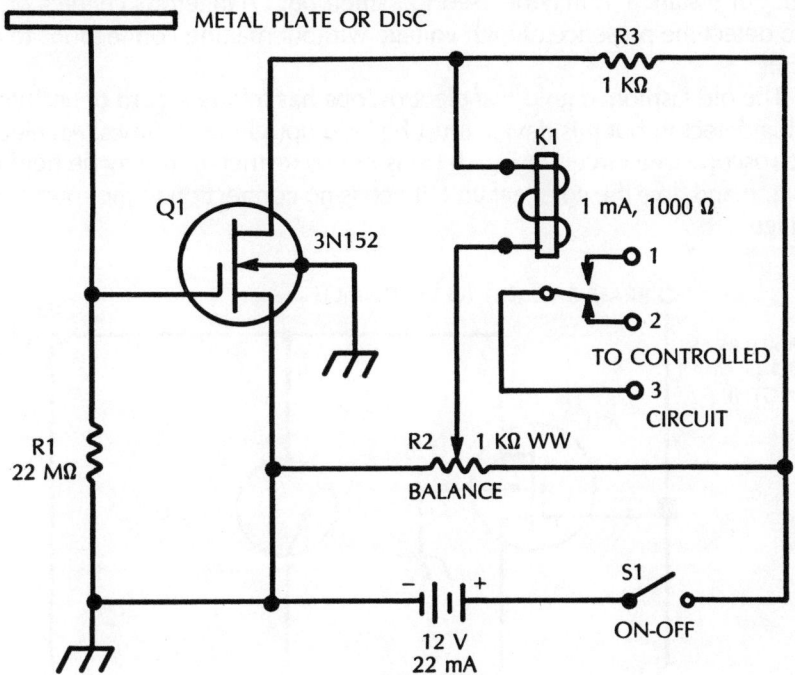

Fig. 3-12. Touch-plate relay.

The MOSFET serves as a simple dc amplifier with a milliampere-type dc relay in its drain circuit. The relay is connected in a four-arm bridge circuit—consisting of resistor R3, the two "halves" of potentiometer R2, and the internal drain-to-source resistance of the MOSFET—which allows the static drain current of the 3N152 to be balanced out of the relay (by adjustment of R2). The relay is a 1 mA, 1000-ohm device (Sigma 5F-1000 or equivalent).

With the pickup disc protected from any nearby bodies, initially set balance control potentiometer R2 to the point at which the relay opens. The circuit will then remain in this balanced, or "zero," condition indefinitely unless the setting of R2 is subsequently disturbed. When the pickup disc is touched, the resulting change in drain current will change the internal resistance of the MOSFET, unbalancing the bridge and causing the relay to close.

Use output terminals 1 and 3 for operations in which the external controlled circuit must be closed by the relay closure; use 2 and 3 for operations in which the external controlled circuit must be opened. When the relay is closed, current drain from the 12-Vdc supply is approximately 22 mA, but this may vary somewhat with individual MOSFETs.

No. 20: Electronic Electroscope

An instrument for detecting the presence of electric charges is indispensable in many situations, in and out of the electronic field. For instance, it may be used to "smell out" dangerous charges of fields, or to detect the presence of high voltage without making connections to a conductor.

The old fashioned gold-leaf electroscope has a long record of usefulness as such a detector, but this device must be held upright at all times. An electronic electroscope (see circuit in Fig. 3-13) is not so restricted; it may be held in any position and, like the gold-leaf unit, it needs no connection to the source of high voltage.

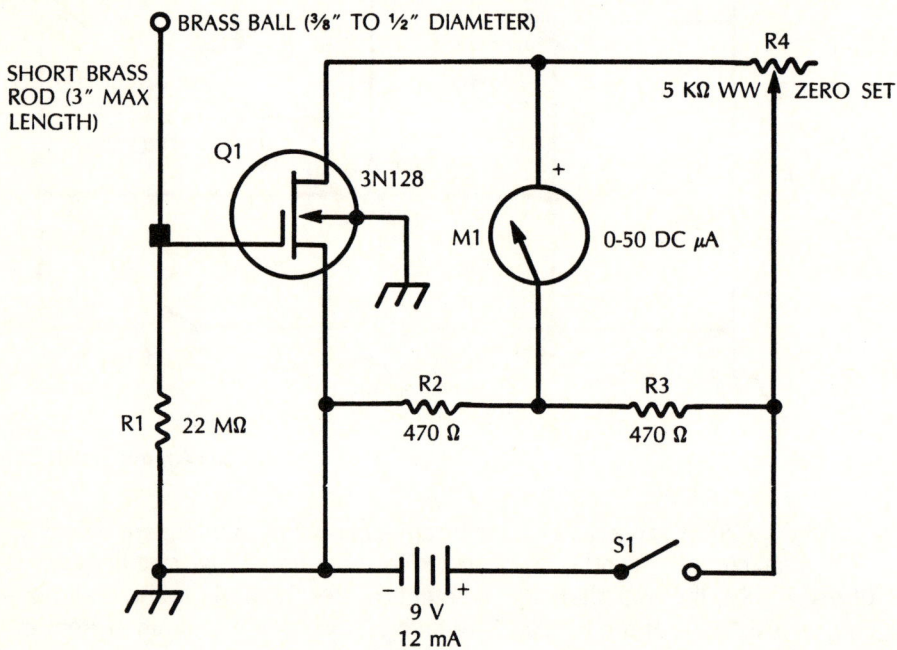

Fig. 3-13.

This arrangement is similar to an electronic dc voltmeter. The gate of the 3N128 MOSFET is 22 megohms above ground and therefore very susceptible to nearby electric charges which are picked up by a polished brass ball on the end of a short rod attached to the gate. The picked-up energy is amplified by the MOSFET, and the latter drives miniature microammeter M1.

As in an electronic voltmeter circuit, the indicating meter is connected in a four-arm bridge circuit in the output of the MOSFET. The arms of this bridge consist of rheostat R4, the internal drain-to-source resistance of the MOSFET, and resistors R2 and R3. With the pickup ball clear of any electric fields and pointed away from the operator's body, potentiometer R4 is set to zero the meter. The device then will respond by an upward deflection of the meter whenever the ball is poked into a field. A substantial deflection is obtained, for example, when the ball is close to the high-voltage wiring in a TV receiver, or when it is confronted with a hair comb that has just been used.

Current drain from the 9-volt battery is approximately 12 mA, but this may vary somewhat with individual MOSFETs. All fixed resistors are ½-watt.

No. 21: Ultrasonic Pickup

Figure 3-14 shows the circuit of an ultrasonic pickup. This device functions as the receptor in an ultrasonic system for remote control, intrusion alarm, code or voice communication, tests and measurements, and similar applications. The pickup can operate directly into an ac circuit; and if a germanium or silicon diode is shunted across the ultrasonic output terminals, it can operate into a dc circuit. The ultrasonic energy is picked up by a 40 kHz transducer, X1.

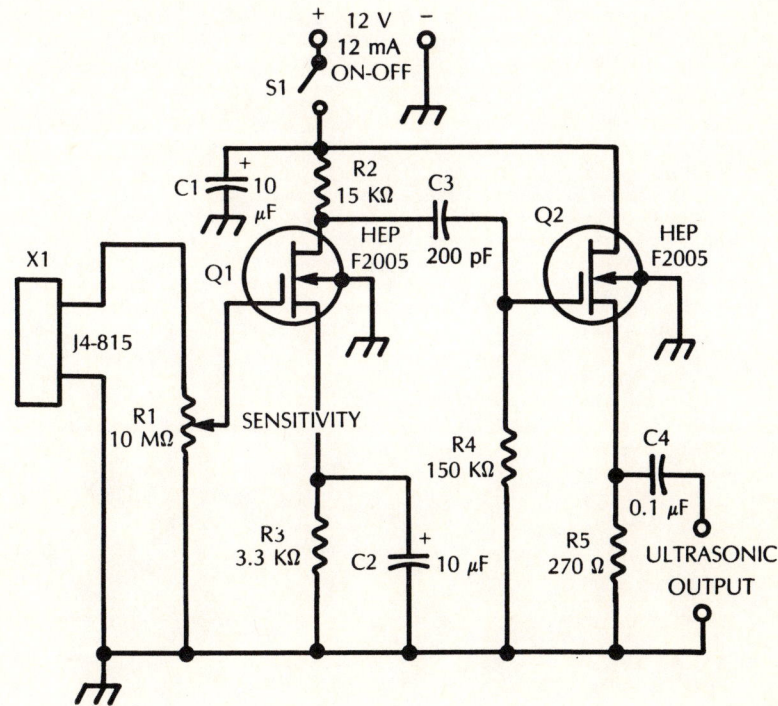

Fig. 3-14.

The circuit consists of a voltage amplifier employing a HEP F2005 single-gate MOSFET (Q1) in a common-source arrangement, and a source follower also employing a HEP F2005 MOSFET (Q2). The voltage gain of the input stage is approximately 30, and that of the output stage approximately 0.7: the gain of the entire pickup circuit thus is approximately 21. The maximum transducer signal, with sensitivity control potentiometer R1 set for maximum sensitivity, is approximately 50 mVrms before peak clipping in the output signal. The corresponding maximum output-signal voltage is approximately 1.05 Vrms. In construction of this device, solid mounting and rigid wiring should be employed to prevent vibration at very high frequencies.

All fixed resistors in this circuit are ½-watt. Electrolytic capacitors C1 and C2 are 25-volt units; capacitors C3 and C4 can be any convenient low-voltage units. The circuit draws approximately 12 mA from the 12-Vdc supply, but this may vary somewhat with individual MOSFETs.

4
Analog ICs

The integrated circuit (IC) consists of a tiny silicon chip which houses a series of components—such as transistors, diodes, resistors, and capacitors—and the interconnecting conductive paths which serve as wiring. These components and paths are formed by a combination of diffusion, photoprinting, and etching. Thus, the chip can contain an entire complicated circuit, such as a multistage amplifier, and the IC will be no larger than a discrete transistor.

The IC, being a complete, miniaturized circuit package, saves much time and labor for the designer and builder of equipment; an entire prewired stage (or full circuit) can be plugged in or removed from a piece of equipment with the ease of handling a transistor or tube. While a large number of ICs contain complicated circuits, some contain only separate components (for example, 2 diodes or 4 transistors) which, being fabricated close together in the same substrate, are excellently matched. Integrated circuits are available in a very large number of types in both linear and digital categories. This chapter will discuss analog ICs.

THEORY

Figure 4-1A shows the structure of a possible integrated circuit; Fig. 4-1B shows the corresponding internal circuit. While this arrangement is not necessarily one that can be found in a manufactured unit, it is true to the nature of the IC and is simple enough for illustrative purposes.

In Fig. 4-1A, n- and p-regions are diffused into the substrate (silicon chip) to form the diode (D1), transistor (Q1), resistor (R1), and capacitor (C1) of the circuit. Next, a silicon dioxide dielectric film is grown on the face of the chip. Then this dielectric is etched away at those places, such as points 4 and 10, where connections must be made to the n- and p-regions. Next, a metal film

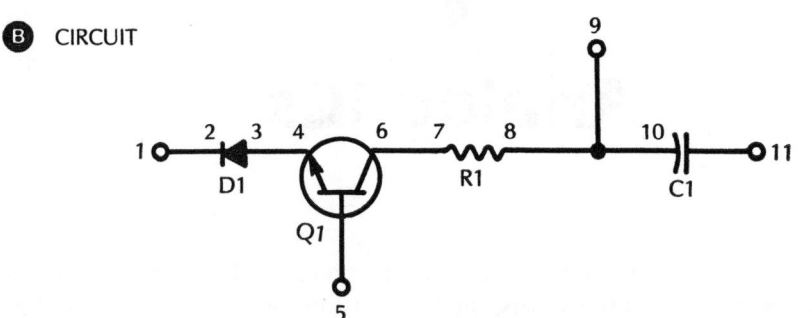

Fig. 4-1. Details of an integrated circuit.

is deposited on top of the dielectric to extend through the etched windows to make contact with the regions and to provide the "wiring" between the components. The matching numbers in Fig. 4-1A and 4-1B allow these contacts and interconnections to be identified.

In this example, the various components are formed in a p-type chip; and since each component has an "outside" n-layer, the resulting pn junction offers very high resistance, being reverse biased, and thus prevents the components from being short-circuited together through the substrate. The resistor consists only of a p-region with connections being made to it at points 7 and 8. One "plate" of the capacitor is the n-region. A connection is made at point 10, and the other plate is the portion "X" of the metal film. This general construction is characteristic of all ICs, although the geometry of a specific manufactured unit may differ considerably from the structure in Fig. 4-1A.

The various functional types of linear ICs are too numerous to list here. However, some familiar examples are dc, af, i-f, rf, video, differential, and operational amplifiers; balanced modulators; product detectors; mixers; and amplifier/discriminators and timers.

Audio Amplifiers

These devices have an entire audio amplifier implemented on a single IC. They can be used as preamplifiers, audio amplifiers, and power amplifiers. Until recently the fidelity and frequency response was not good enough to be considered useful in expensive stereo applications, but new designs bring these chips within range of high-end audio equipment. Some audio ICs have a preamplifier and a power amplifier on the same chip. The MC1306 is an example of such a device.

Op Amps

Operational amplifiers are extremely stable linear amplifiers. They can be used in a great number of applications. They are very often used in analog computational circuits. Functions like differentiation, integration, summation, or division are easily accomplished. They can also be used as comparators.

Many op amps use a split supply ($-V_{CC}$ and $+V_{CC}$). The 741 op amp is an example. Refer to Fig. 4-2 for the circuitry that is on the 741 chip.

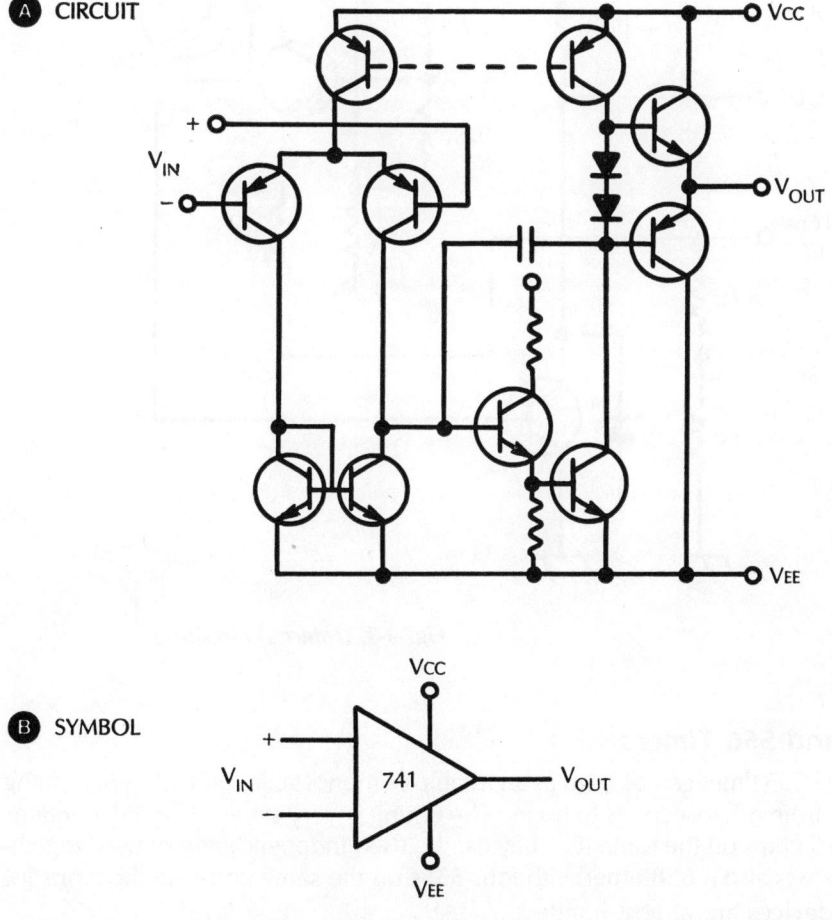

Fig. 4-2. Internal circuitry and schematic symbol for an op amp.

3909 Flasher

The 3909 Flasher IC is an extremely versatile chip. It can be used as an oscillator, an amplifier, a timer, or an AM detector. It can even be used to perform digital functions. Another nice feature of this IC is its ability to operate from a 1.5 volt cell for extended periods.

The internal circuitry of the 3909 is shown in Fig. 4-3. The 6 kΩ and the 3 kΩ resistors are the internal RC timing resistors. Q3 is the power transistor with a 150 mA capability. It can drive a permanent magnet loudspeaker directly.

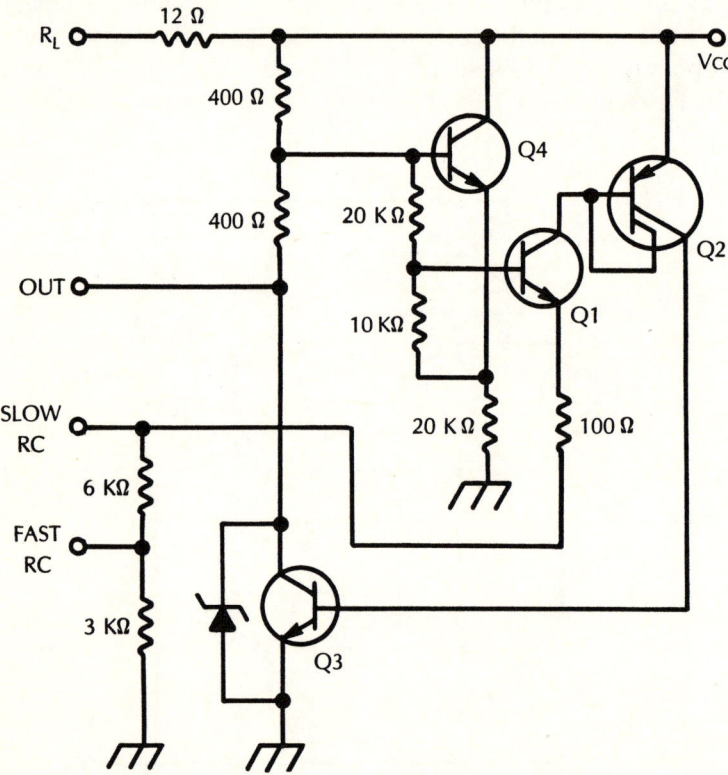

Fig. 4-3. (Internal circuitry)

555 and 556 Timers

The 555 timer can be used as an astable or monostable multivibrator. Timing can be from microseconds to hours. The timing is very stable. The 556 contains two 555 chips on the same IC. They can be used independently or used together. There is also a 558 timer with four 555s on the same chip. Applications for these devices are almost limitless.

Complex Sound Generator

The SN94281 IC combines both linear and digital circuitry on the same chip. A variety of sound effects, noises, and tones can be created by varying the external components and the digital inputs. This particular Sound Generator has an on-board 125 mW audio amplifier. It can drive a small 8 Ω speaker with sufficient volume. Some versions of this IC are available without the audio amplifier. Applications with these ICs will require an external transistor or IC amplifier.

No. 22: Audio Amplifier

One of the attractions of the IC to experimenters and some audiophiles is the ability of some ICs to supply a complete audio amplifier in a small transistor-type case. An amplifier of this type is shown in Fig. 4-4. Here, the CA3020 integrated circuit contains three intermediate stages and a push-pull, class-B output stage. The only "outboard" components required are four capacitors, two fixed resistors, one potentiometer, and one output transformer. The circuit operates from a single battery.

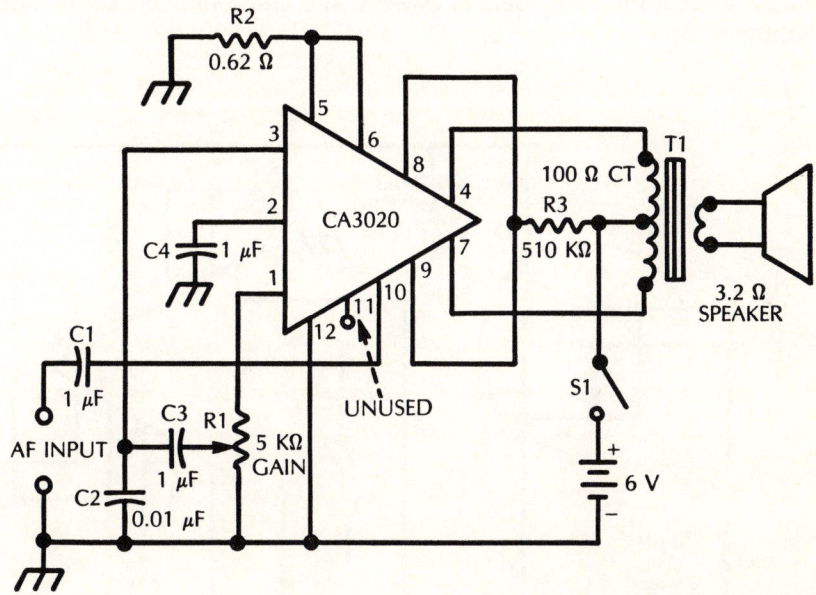

Fig. 4-4. Complete audio amplifier.

The amplifier will operate from microphones, phono pickups, photocells, and other transducers, and delivers ¼-watt output. Output transformer T1 is a miniature unit, such as Argonne AR-174, having a 100- or 125-ohm center-tapped primary winding and 3.2-ohm secondary, to match the 3.2-ohm speaker to the IC.

The class-B output stage uses most of the dc input power; thus, the idling current drain from the 6-volt battery is approximately 4 mA, and the maximum-signal current is 77 mA. At a battery voltage of 9 volts, the current drains are somewhat higher, and the power output is approximately 0.3 watt. With either 6- or 9-volts input, however, the IC must be supplied with a suitable heat sink.

The amplifier exhibits good gain. An input of 50 mV rms results in full ¼-watt output with low distortion when gain control R1 is set for maximum gain. This operation allows the amplifier to function as the complete af channel of a receiver, monitor, or test instrument. It also serves as its own integral audio amplifier.

For best stability, resistors R2 and R3 should be 1 watt. Capacitors C1, C3, and C4 can be miniature tantalum electrolytics, if maximum compactness is desired; otherwise any convenient low-voltage nonelectrolytics may be used. Capacitor C2 may be paper, ceramic, or mica. Size-D flashlight cells give reasonably good service for short-interval operation, but miniature transistor-type batteries are short lived in this circuit.

Other ICs are also available for complete-amplifier operation. One example is type LM380. This type delivers 2½ watts (total harmonic distortion = 0.2%) with a 22-Vdc supply, and operates directly into an 8-ohm speaker without an output transformer.

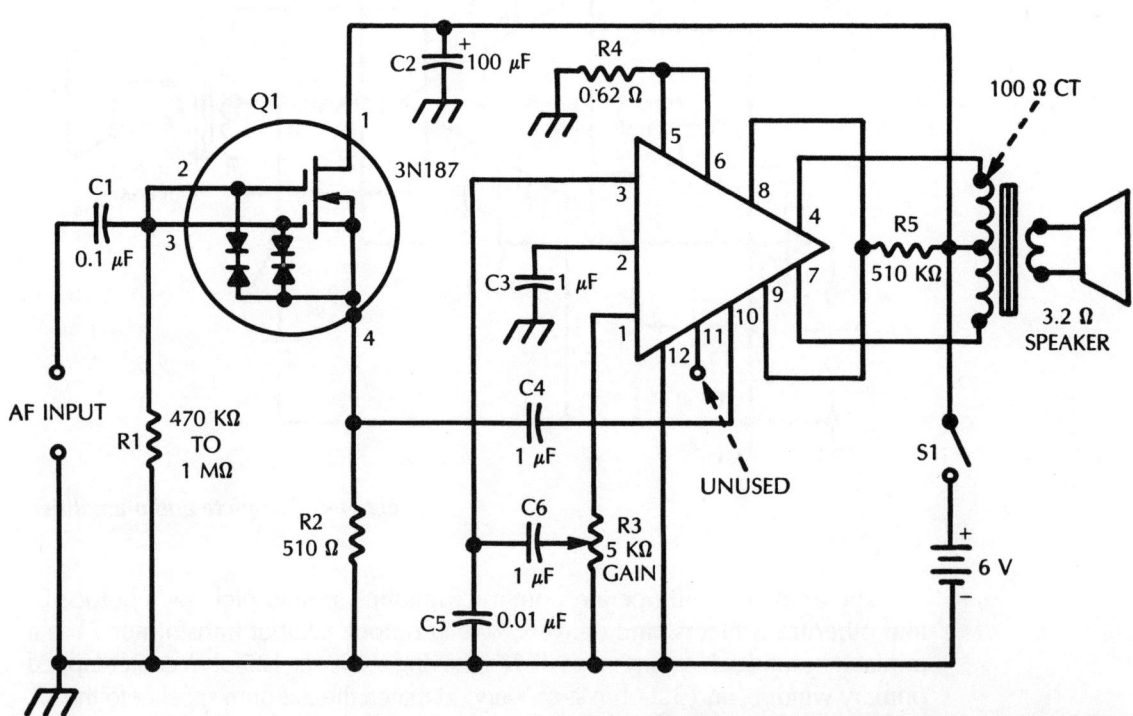

Fig. 4-5. Audio power amplifier with MOSFET input.

No. 23: Audio Amplifier with MOSFET Input

The versatility of the complete audio amplifier described in the preceding section can be widened by providing it with a high-impedance input stage. Figure 4-5 shows how a 3N87 MOSFET is employed as a simple source follower to provide this high impedance. (See Chapter 3 for MOSFET theory and application.)

The voltage gain of the input stage is approximately 0.51, and this requires that the input signal amplitude be 98 mV rms for a full ¼-watt output into the speaker. For this circuit, the zero-signal current from the 6-volt battery is 14 mA, and the maximum-signal current is 87 mA. These are approximate figures and will vary somewhat with individual ICs and MOSFETs.

Five outboard components are required for the MOSFET: Capacitor C2 is a 25-volt electrolytic unit, while C1 may be any convenient low-voltage unit. Resistors R1 and R2 are ½-watt. All outboard components in the IC portion of the circuit correspond to those in the same positions in the basic IC circuit, Fig. 4-4.

For ease of handling and installation, a gate-protected MOSFET has been chosen; however, a conventional MOSFET also can be used if the experimenter is experienced in handling the unprotected type (see, for example, the source follower in Fig. 3-7).

No. 24: 2½-Watt Intercom

Figure 4-6 shows the circuit of a miniature intercom employ-, ing a type LM380 integrated circuit as the complete amplifier. This IC delivers 2½-watts output at full gain and operates directly into an 8-ohm speaker, without a coupling transformer. A small speaker is used at both stations as microphone and reproducer, depending upon the position of the talk-listen switch, S1 and S3. The two stations are interconnected by means of a 3-wire, shielded cable.

When a speaker is acting as a microphone (S1 or S3 in the talk position), it is coupled to the input of the IC by T1, a small 4000:8 ohm output transformer (such as Argonne AR-134) connected backward. The other speaker then is automatically connected to the output of the IC and serves as a reproducer. The 2½-watt output will be adequate in all except very noisy places.

Construction of the intercom should be entirely straightforward. Take special precautions to keep the input and output circuits of the IC as isolated as well as practicable, to prevent unwanted electrical feedback, and to provide the IC with a heat sink. Aside from the input coupling transformer, the only outboard components are the 1-megohm rheostat (R1) and 500 μF output capacitor (C1).

Current drain from the 22.5-Vdc source is approximately 25 mA zero-signal, and 220 mA maximum signal. These currents may vary somewhat with individual ICs.

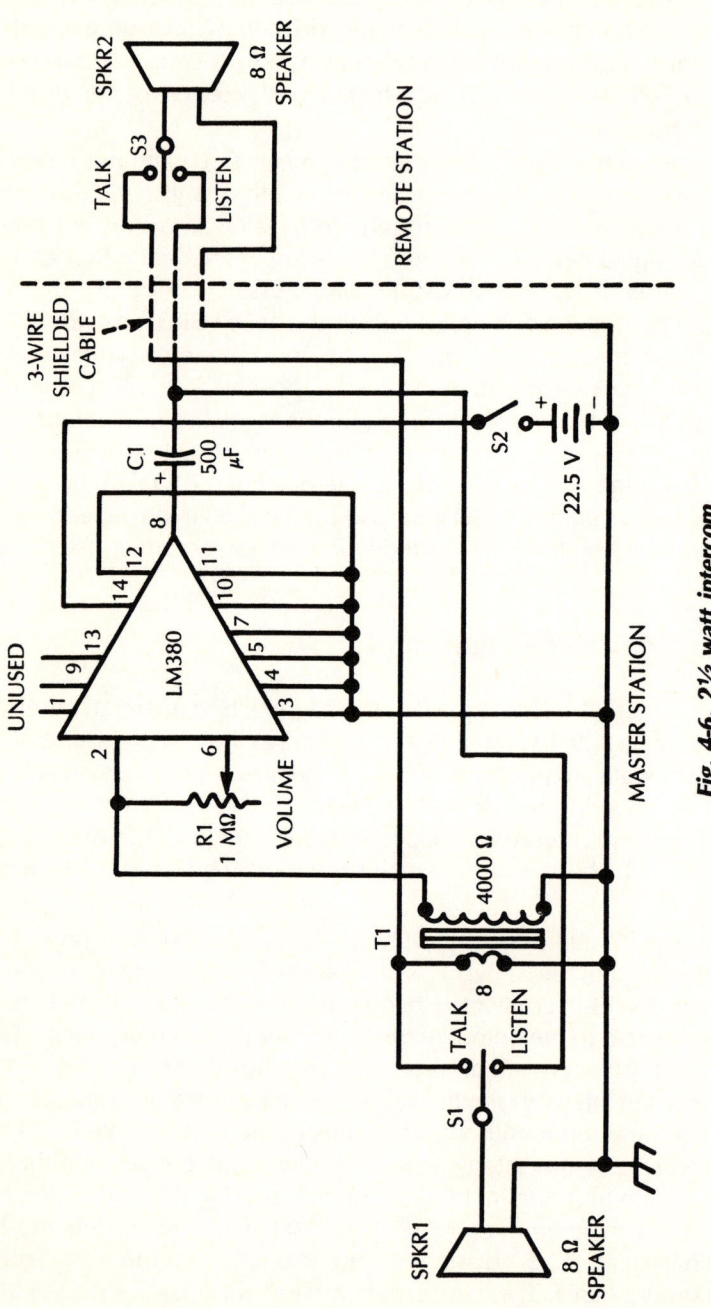

Fig. 4-6. 2½ watt intercom.

No. 25: Low-Resistance dc Milliammeter

Low resistance is desired in a current meter, in order to minimize the voltage drop introduced by the meter into a circuit where it is inserted. The more closely the internal resistance of the meter approaches zero, the less the instrument upsets the circuit. Even though the resistance of many current meters is low, it still is too high for some applications unless it is painstakingly taken into account in all measurements and calculations. For example, the common resistance of the useful 0-1 dc milliammeter is 55 ohms, corresponding to a voltage drop of 55 mV.

Figure 4-7 shows the circuit of an electronic dc milliammeter giving a full-scale deflection of 1 mA and having a resistance of only 2.7 ohms. In this circuit, the type µA776 integrated circuit functions as a dc voltage amplifier with a voltage gain of approximately 40, and boosts the 2.7 mV developed by the 1 mA input current across resistor R1 to the 100 mV required to deflect the 0-50 dc microammeter (M1) to full scale. If this microammeter—whose internal resistance is 2000 ohms—were simply shunted to convert it into a 0-1 milliammeter, a shunt resistance of 105 ohms would be required, and the resulting instrument resistance would be 99.8 ohms, approximately 37 times the input resistance of the electronic circuit.

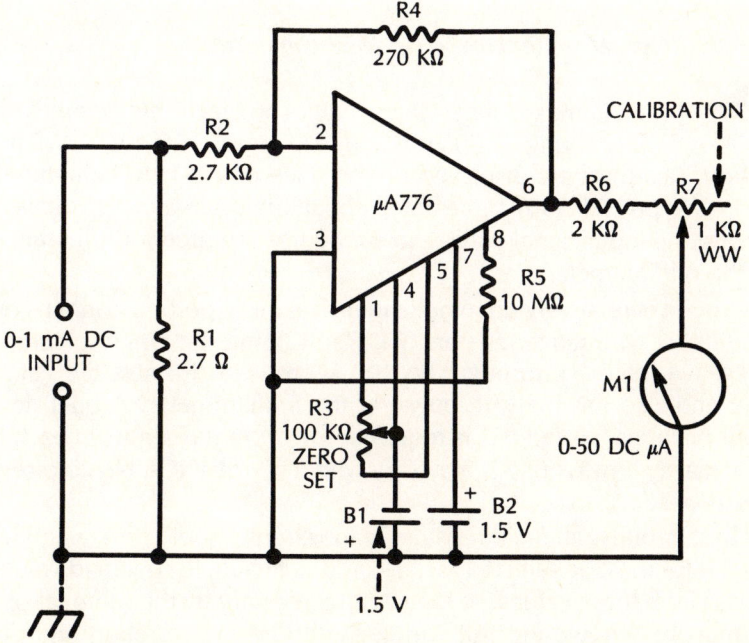

Fig. 4-7.

In this circuit, resistors R1, R2, and R4 must be selected to have resistance as close as practicable to the specified values. If precision instrument resistors are available to the builder, R1 can be 2 ohms, R2 2 kΩ, and R4 200 kΩ; the input resistance then will be 2 ohms. Meter M1 is initially set to zero by adjustment of potentiometer R3 which adjusts the offset of the IC. Rheostat R7 allows the meter to be set for exact full-scale deflection when an accurately known 1 mA current is applied to the dc input terminals. For best stability, resistors R1, R2, and R4 should be 1-watt each. Resistors R5 and R6 may be ½-watt. Rheostat R7 must be wirewound.

Initial adjustment of the instrument is simple:

1. With the dc input terminals open, set rheostat R7 to its zero-resistance position and set potentiometer R3 to zero the microammeter.
2. Apply an accurately known 1 mA direct current to the dc input terminals.
3. Adjust rheostat R7 for exact full-scale deflection of the meter.
4. The instrument response is linear, and may be checked by applying successively lower currents (such as ½ mA, ¼ mA, etc.) to the dc input terminals. A special 0-1 mA card may be drawn for the meter, or the 0-50 µA card may be retained and the readings in mA interpreted from the µA deflections.

An added advantage of this circuit is its low power demand. Size-D flashlight cells at B1 and B2 will give an adequate shelf life, so no on-off switch is required.

No. 26: Electronic dc Millivoltmeter

Figure 4-8 shows the circuit of an electronic dc millivoltmeter covering 0-1000 mV in three ranges: 0-10 mV, 0-100 mV, and 0-1000 mV. Readings are displayed on the scale of a 0-1 dc milliammeter. In this circuit, a type HEP S3001 integrated circuit functions as a dc voltage amplifier to boost the input-signal voltage to the 55 mV required for full-scale deflection of the milliammeter.

The circuit is relatively uncomplicated. The input portion contains only the three multiplier-type range resistors (R1, R2, R3) and the range selector switch (S1). Negative feedback through the 100 kΩ resistor, R4 sets the gain of the amplifier and also linearizes response of the milliammeter. A dual dc supply (22.5-volt batteries B1 and B2) is required; the current drain from each battery is approximately 5 mA, but this may vary with individual ICs. No zero-set potentiometer is needed.

For best stability, all fixed resistors in the circuit should be 1-watt. Resistors R1, R2, and R3 must be selected as close as practicable to specified values (they may depart from these values, so long as all three vary by the same ratio). Rheostat R6 must be wirewound. All wiring should be kept as short and direct as possible. Unused terminals of the IC may be left floating.

To calibrate the instrument initially:

1. Set switch S1 to its 1000 mV position.

2. Set calibration rheostat R6 to its zero-resistance position. Close switch S2.
3. Apply an accurately known 1-Vdc to the dc input terminals, observing correct polarity.
4. Adjust rheostat R6 for exact full-scale deflection of milliammeter M1.
5. Operation of the circuit is linear, and this may be checked by applying successively lower voltages (with S1 still set to the 1000 mV range), such as ½-volt, ¼-volt, etc., while observing deflection of the meter.
6. If resistors R1, R2, and R3 are accurate, this one calibration will serve also for the 100 mV and 10 mV ranges which will then be automatically in calibration.

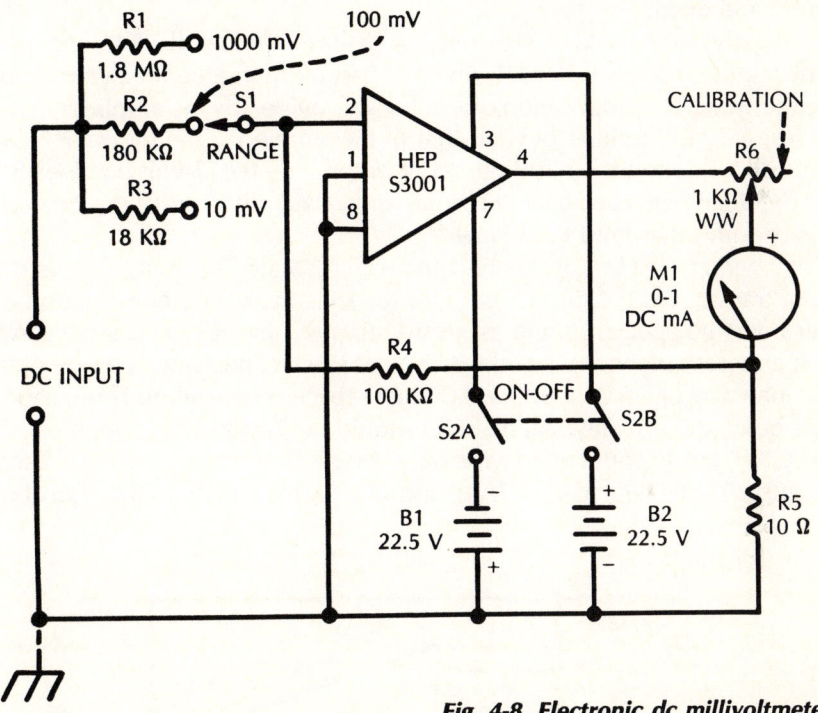

Fig. 4-8. Electronic dc millivoltmeter.

No. 27: Low-Pass Active Filter

Figure 4-9 shows the circuit of a low-pass active filter employing a type 741 operational-amplifier IC. This circuit provides a cutoff frequency of 500 Hz, but may be adapted to other frequencies by suitably changing the values of R1, R2, C1, and C2. (For other frequencies, keep R1 equal to R2, and C1 equal to C2. The cutoff frequency f_c equals:

$$f_c = 10^6/(2\pi RC)$$

where f_c is in Hz, R in ohms, and C in μF.) Beyond the cutoff frequency, the response of the circuit falls off at the rate of 12 dB per octave. Overall voltage gain is approximately 1.59.

Negative feedback is provided by voltage divider R3-R4, and positive feedback through capacitor C1 which is part of the frequency-determining RC filter network. This application of feedback, provided by the amplifier, sharpens the response of the filter beyond that of the simple RC circuit (R1-R2-C1-C2).

A dual dc supply is required, represented here by 12-volt batteries B1 and B2. The drain from each battery is approximately 3 mA, but this can be expected to vary somewhat with individual ICs.

All wiring must be kept as short and well separated as practicable, to prevent signal transfer around the IC; in other respects, however, construction of this filter offers no special problems. Resistances R1 and R2 must be accurate, as must also capacitances C1 and C2. The 0.032 μF capacitance can be obtained by connecting one 0.03 and one 0.002 μF capacitor in parallel. These capacitors must be of good quality. All resistors should be 1-watt. Dc coupling is shown in Fig. 4-9, but ac coupling may be obtained by connecting one 10 μF capacitor in series with the high af input lead, and another in series with the high af output lead.

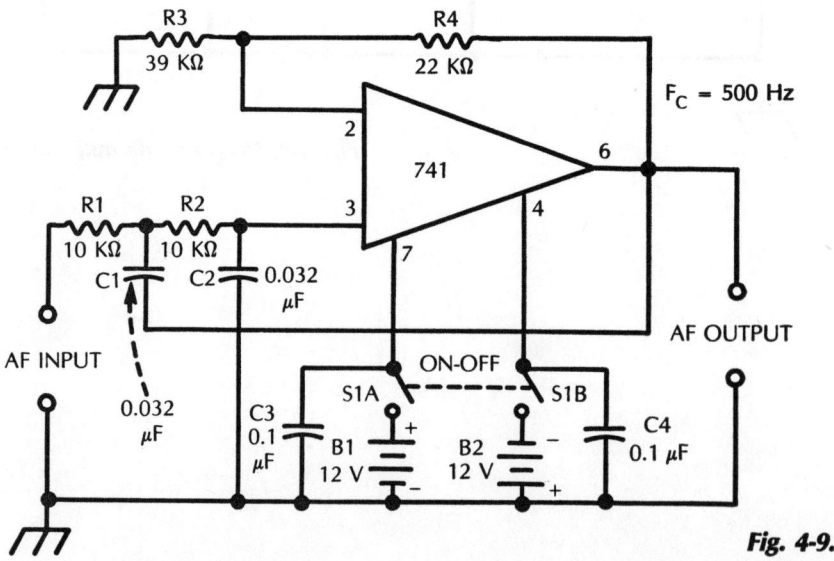

Fig. 4-9.

No. 28: High-Pass Active Filter

Figure 4-10 shows the circuit of a high-pass active filter employing a type 741 operational-amplifier IC. This circuit, like the low-pass filter described in the preceding section, provides a cutoff frequency of 500 Hz, but may be adapted to other frequencies by suitably changing the values of R1, R2, C1, and C2. For other frequencies, keep R1 equal to R2, and C1 equal to C2. The cutoff frequency f_c equals:

$$f_c = 10^6/(2\pi RC)$$

where f_c is in Hz, R in ohms, and C in μF.

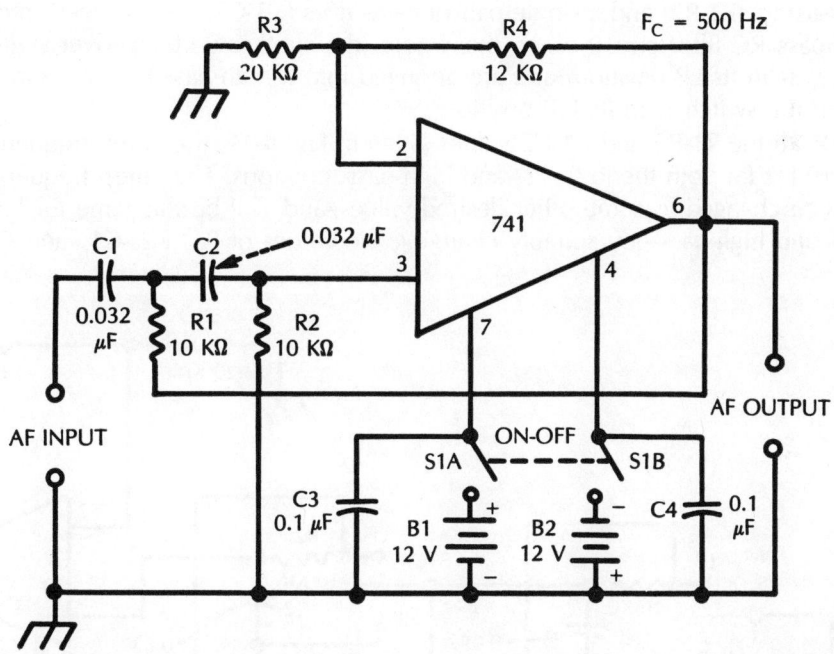

Fig. 4-10.

Below this cutoff frequency, the response of the circuit rises at the rate of 12 dB per octave. Overall voltage gain is approximately 1.59.

Negative feedback is provided by voltage divider R3-R4, and positive feedback through resistor R1 which is part of the frequency-determining RC filter network. This application of feedback, provided by the amplifier, sharpens the response of the filter beyond that or the simple RC circuit (R1-R2-C1-C2).

A dual dc supply is required, represented here by batteries B1 and B2. The drain from each battery is approximately 3 mA, but this can be expected to vary somewhat with individual ICs.

All wiring must be kept as short and well separated as practicable, to prevent signal transfer around the IC; in other respects, however, construction of this filter offers no special problems. Resistances R1 and R2 must be accurate, as must also capacitances C1 and C2. The 0.032 µF capacitance can be obtained by connecting one 0.03 and one 0.002 µF capacitor in parallel. These capacitors must be of good quality. All resistors should be 1-watt. The input is ac coupled by the first filter capacitor, C1, but the output is dc coupled. If desired, the output can be ac coupled by inserting a 10 µF capacitor in the high af output lead.

No. 29: Combination Active Filter

The active filter circuit shown in Fig. 4-11 is, in effect, a combination of a low-pass filter and high-pass filter. A single pair of resistors (R1-R2) and a single pair of capacitors (C1-C2) are arranged into a low-pass RC filter circuit when the 4-pole, double-throw changeover switch ($S1_{A-D}$) is in its LP position, and are arranged into a high-pass RC filter circuit when the switch is in its HP position.

With the R1-R2 and C1-C2 values given in Fig. 4-11, the cutoff frequency is 500 Hz for both the low-pass and high-pass functions. The cutoff frequency may be changed to some other desired value—and will be the same for low-pass and high-pass—by suitably changing the values of R1, R2, C1, and C2.

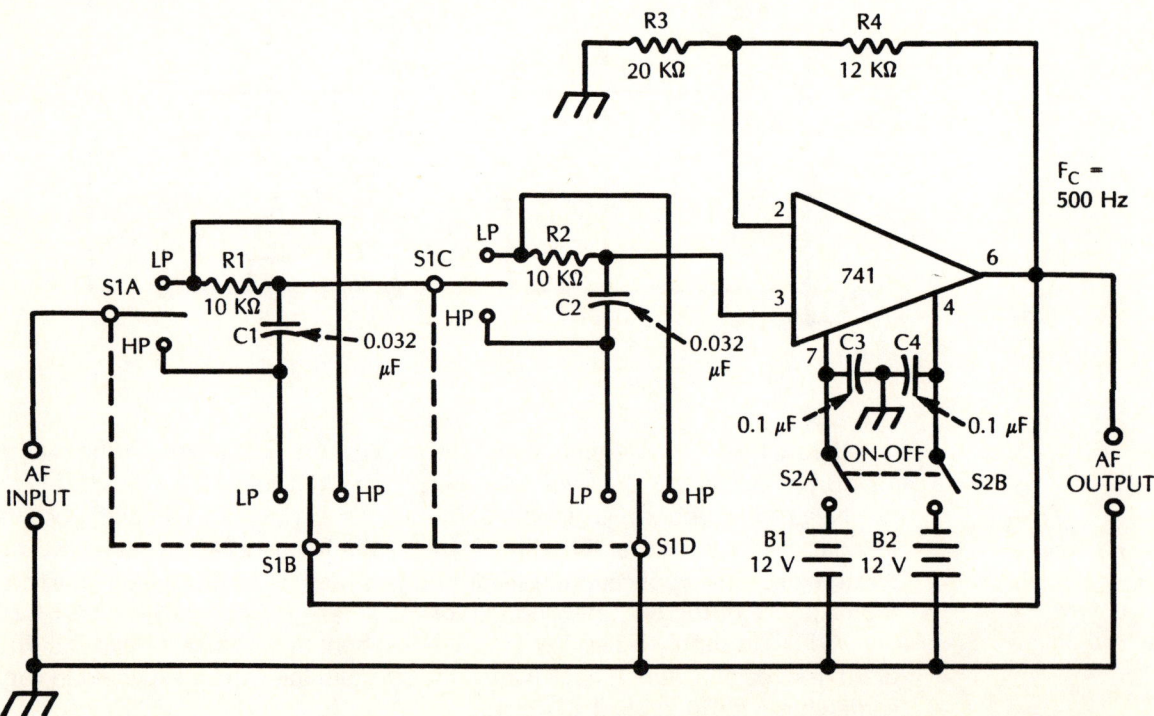

Fig. 4-11. Combination high-pass and low-pass active filter.

For other frequencies, keep R1 equal to R2, and C1 equal to C2. The cutoff frequency f_c equals:

$$f_c = 10^6/(2\pi RC)$$

where f_c is in Hz, R in ohms, and C in µF.

Above this frequency, in the low-pass function, the response of the circuit falls off at the rate of 12 dB per octave; below this frequency, in the high-pass function, the response rises at the rate of 12 dB per octave.

The general characteristics and construction of this filter are the same as those separately described for the low-pass filter and for the high-pass filter. See earlier sections of this chapter for this information. The combination filter has dc-coupled output, but the input is inherently dc coupled on low-pass and ac coupled on high-pass. If ac coupling is desired for both high-pass and low-pass and for both input and output, insert one 10 µF capacitor into the high af input lead, and another 10 µF into the high af output lead.

No. 30: Bandpass Active Filter

Figure 4-12A shows the circuit of a bandpass active filter of the sharp-peak type employing a type 741 operational-amplifier IC. A parallel-T RC network (R3-R4-R5-C2-C3-C4) in the negative-feedback loop determines the peak frequency of this filter. This is a null network, but it makes a pass amplifier out of the circuit, as it cancels gain of the amplifier on all frequencies except the null frequency. Positive feedback is adjustable by means of rheostat R7, and serves to sharpen the response of the circuit, thus effectively increasing the Q of the parallel-T network.

Figure 4-12B shows the response of the circuit: The solid curve depicts minimum selectivity (low Q) when positive feedback is minimum (R7 set to maximum resistance), whereas the dotted curve shows maximum selectivity (high Q) when positive feedback is maximum (R7 set to low or zero resistance). Voltage gain of the circuit at the peak of the response is approximately 100.

The resistance and capacitance values in the parallel-T network have been chosen for a filter peak of 1000 Hz. The peak frequency may be changed to some other desired value, however, by suitably changing the values of C2, C3, C4, R3, R4, and R5.

For other frequencies, keep R3 = R5 = R4, and C2 = C3 = ½C4. The peak frequency f equals:

$$f = 10^6/(2\pi R3C2)$$

where f is in Hz, R3 in ohms, and C2 in µF.

A dual dc supply is required, represented here by 15-volt batteries B1 and B2. The drain from each battery is approximately 4 mA, but this can be expected to vary somewhat with individual ICs.

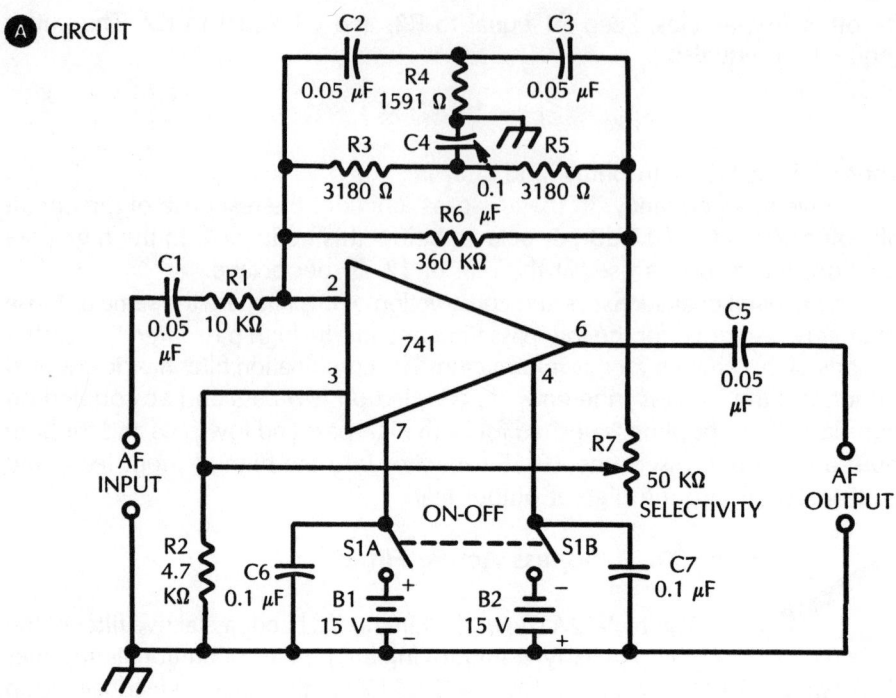

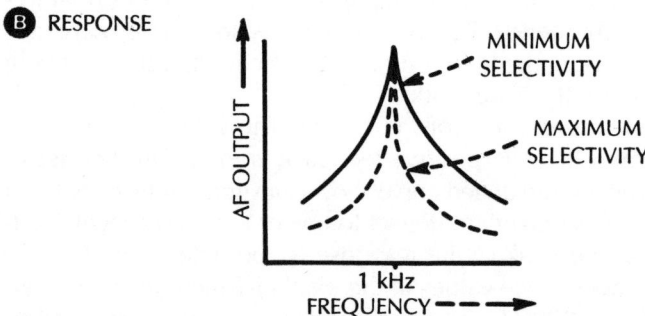

Fig. 4-12. Bandpass active filter.

All wiring must be kept as short and well separated as possible, to prevent signal transmission around the IC; in other respects, however, construction of the filter offers no special problems. Resistances R3, R4, and R5 must be accurate, as must capacitances C2, C3, and C4. The off-standard values for R3 and R5 can be obtained by connecting one 3000- and one 180-ohm resistor in series, and for R4 by connecting one 1500- and one 91-ohm resistor in series. Capacitors C2, C3, and C4 must be of excellent quality.

It is interesting to note that the filter becomes a sine-wave oscillator when the positive feedback is advanced sufficiently (that is, when selectivity control R7 is set for excessive regeneration).

No. 31: Active Notch Filter

Figure 4-13A shows the circuit of a bandstop active filter of the sharp-notch type, employing a type 741 operations-amplifier IC. A parallel-T RC network (R1-R2-R3-C1-C2-C3) in the input circuit determines the notch frequency of this filter. This is a null network. Positive feedback is introduced at the common point of this network (junction of C3 and R3) to sharpen the response of the parallel-T. Figure 4-13B shows typical response.

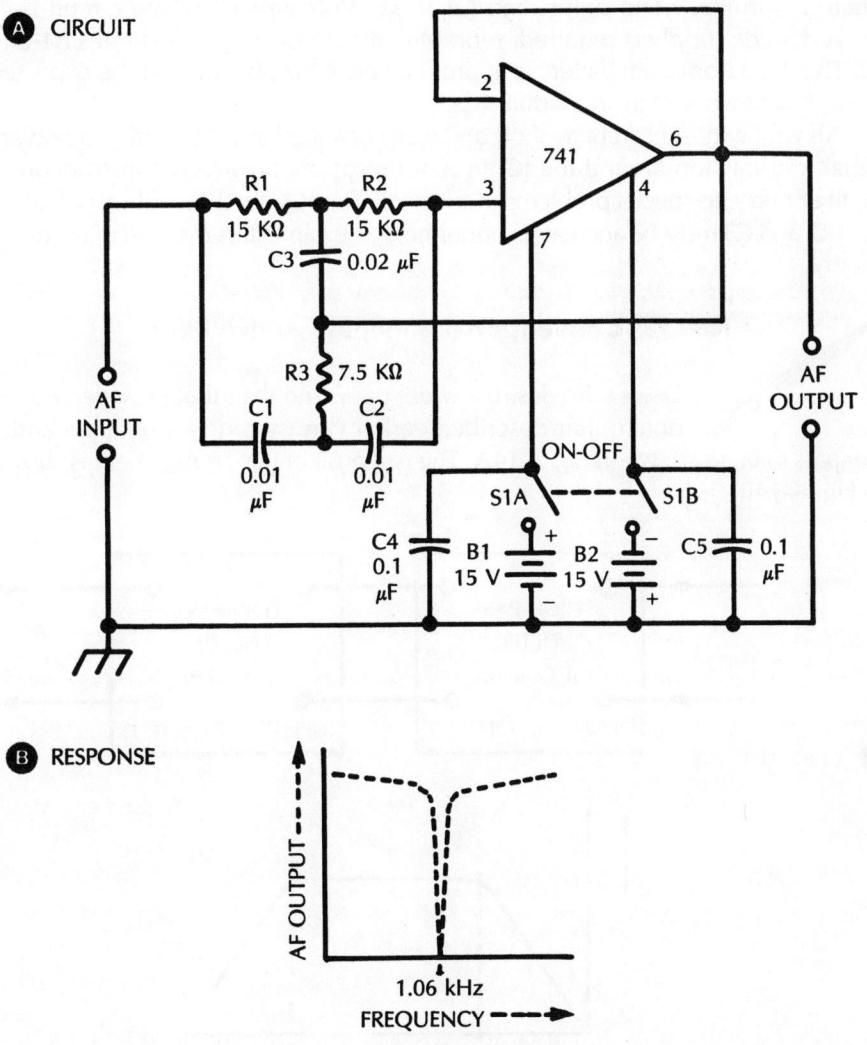

Fig. 4-13. *Active bandstop filter.*

The resistance and capacitance values in the parallel-T network are stock, and the resulting notch frequency is 1061 Hz. For exactly 100 Hz, R1 = R2 = 15,915 ohms, R3 = 7957 ohms; and the capacitances remain the same as shown in the circuit. The notch frequency may be changed to any other desired value, by suitably changing the resistance or capacitance values.

For other frequencies, keep C1 = C2 = ½C3, and R1 = R2 = 2R3. The notch frequency f equals:

$$f = 10^6/(2\pi R1 C1)$$

where f is in Hz, R1 in ohms, and C1 in μF. Voltage gain of the circuit is 1.

A dual dc supply is required, represented here by 15-volt batteries B1 and B2. The drain from each battery is approximately 4 mA, but this can be expected to vary somewhat with individual ICs.

All wiring must be kept as short and well separated as practicable, to prevent signal transmission around the IC; in other respects, however, construction of the filter offers no special problems. Resistances R1, R2, and R3, and capacitances C1, C2, and C3 must be accurate. Furthermore, the capacitors must be of excellent quality.

No. 32: Conventional Bandpass Active Filter

Users who desire a wider passband than that provided by the notch filter described earlier can cascade a high-pass and a low-pass filter, as shown in Fig. 4-14A. The response of this arrangement is shown in Fig. 4-14B.

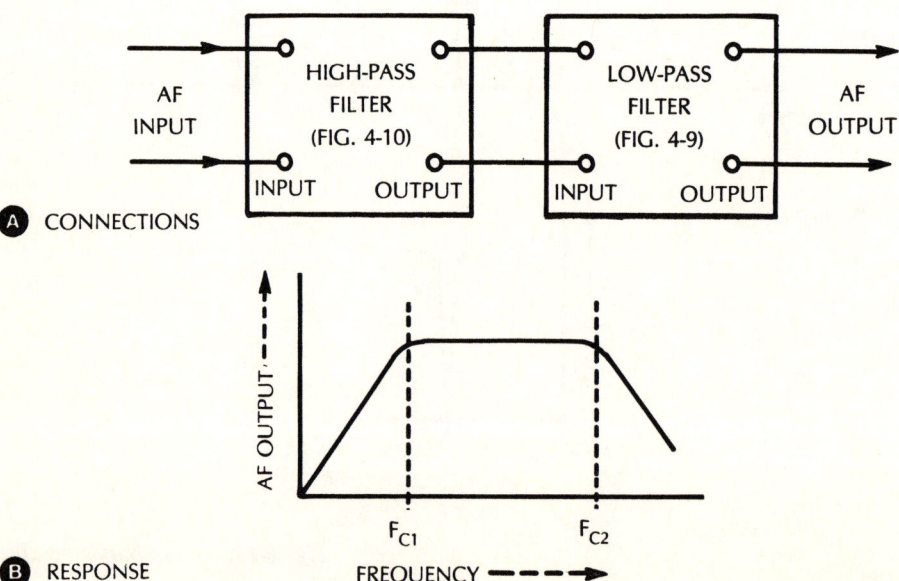

Fig. 4-14.

In Fig. 4-14B, f_{C1} is the cutoff frequency of the high-pass unit, and f_{C2} is the cutoff frequency of the low-pass unit. The high-pass and low-pass units may be designed for the desired f_{C1} and f_{C2} frequencies. The passband of the complete bandpass filter then is equal to $f_{C2} - f_{C1}$. Below f_{C1}, the response of the bandpass filter rises at the rate of 12 dB per octave; and beyond f_{C2}, the response falls at a rate of 12 dB per octave.

No. 33: Signal Tracer

An untuned signal tracer remains an invaluable electronic troubleshooting aid. Such a device should provide a reasonable amount of signal sensitivity and visual, as well as aural indications. The conventional signal tracer embodies a high-gain audio amplifier with a power output stage, provides high input impedance, and works from a straight test probe (for audio testing) or a demodulator probe (for modulated-rf testing). Battery operation gives complete isolation from the power line.

Figure 4-15 shows the circuit of an efficient signal tracer that can be built to small dimensions. A type LM380 integrated circuit supplies the high-gain audio amplifier with power-output stage. A source follower, employing a 3N128 MOSFET, connected ahead of the IC provides the very high input impedance. The shielded jack, J1, receives either the straight test probe or the demodulator probe.

When switch S2 is in its aural position, the amplifier drives the 8-ohm speaker; no output transformer is needed. When S2 is in its visual position, the 8.2-ohm resistor, R4, substitutes for the speaker as the amplifier load, and a rectifier-type meter (C5-D1-D2-R5-M1) indicates the output-signal voltage developed across this resistor. Rheostat R5 may be set initially to confine the deflection to full scale when gain control R3 is set for maximum gain (highest tracer sensitivity); R5 then needs no subsequent adjustment, except for periodic realignment to correct aging effects.

Since the IC contains a class-B output stage, the current drawn from the 15-Vdc source (battery B1), varies with the input signal: the zero-signal level is 10 mA, and the maximum-signal level 150 mA.

The simplicity of the circuit insures troublefree construction. Except for the care needed in handling the MOSFET (see precautions in Chapter 1), no special problems are introduced. Fixed resistors R1 and R2 are ½-watt; R4 is 5 watts. Electrolytic capacitors C2 and C4 are 25-volt units; capacitors C1, C3, and C5 may be any convenient low-voltage units, except C5 should not be electrolytic.

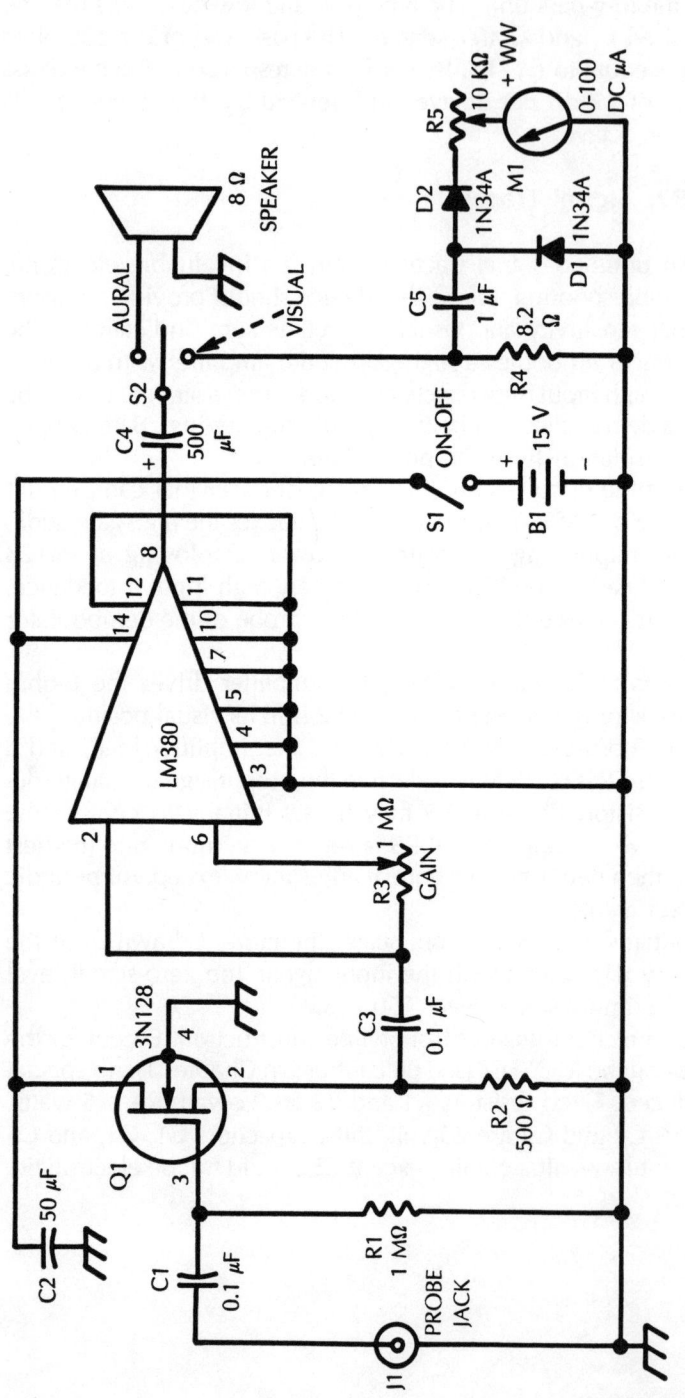

Fig. 4-15. Signal tracer.

No. 34: Code-Practice Oscillator

The resistance-capacitance af oscillator shown in Fig. 4-16 delivers enough audio output—approximately 2-watts—to an 8-ohm speaker to serve a group of listeners. The type LM380 integrated circuit operates as a phase-shift oscillator with a phase-lag-type frequency-determining network (R1-R2-R3-C1-C2-C3) inserted into its positive-feedback loop.

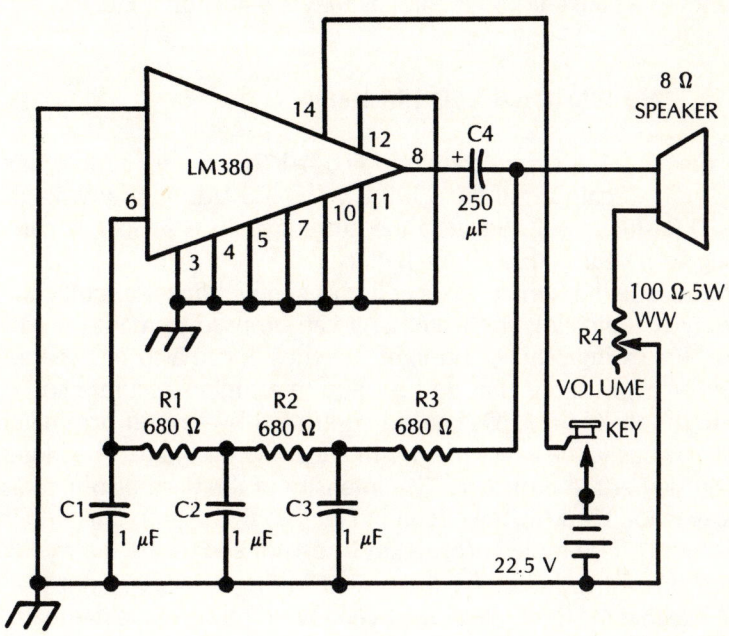

Fig. 4-16.

With the stock resistance and capacitance values in the network, the frequency is 406 Hz. For exactly 400 Hz, with the 1 µF capacitors shown, the resistors would each need to be 688.7 ohms. The frequency of approximately 400 Hz was chosen for its easy "readability." However, the frequency of the oscillator can be changed to any other desired value by suitably changing the resistance or capacitance values in the phase-shift network. For any frequency, keep R1 = R2 = R3, and C1 = C2 = C3, the oscillation frequency f equals:

$$f = 10^6/(3.63\ f\ C)$$

and

$$C = 10^6/(3.63\ f\ R).$$

The R and C values thus obtained will always produce the required total phase shift of 180 degrees. Resistors R1, R2, and R3 must be accurate, as must also capacitors C1, C2, and C3.

When the key is down, the circuit draws approximately 220 mA from the 22.5-volt dc supply, represented here by battery B1. The IC should be provided with a heat sink.

Resistors R1, R2, and R3 should be 1-watt. The 100-ohm volume control rheostat, R4, must be a 5-watt unit. Capacitors C1, C2, and C3 may be any convenient low-voltage units, so long as they are accurate. Electrolytic capacitor C4 should be a 25-volt unit.

No. 35: Dual LED Flasher

The circuit shown in Fig. 4-17A is a very simple, but practical, use of the 3909 IC. The 1.5 Vdc supply voltage can be any C or D flashlight cell. Because the current drain is so low, it can operate for as long as a year with a single battery.

With the long life from a single cell, this simple blinker circuit can be used for a variety of indicating applications. It can be used to mark the location of flashlights, fire extinguishers, and light switches. It can also be used as a pulse generator for applications that do not require a highly accurate source.

The flash rate of the LEDs can be controlled by several parameters. First, and most obviously, the value of C1 will affect the flash rate. The larger the capacitor the slower the flash rate. The intensity of the flash also increases with a larger capacitor. By using the circuit in Fig. 4-17B, the flash rate with the same size capacitor can be tripled. The 1000 Ω resistor shunts the internal 6 kΩ and 3 kΩ time-constant resistors. Try the circuit in Fig. 4-17B without the 1000 Ω resistor. Note that the flash-rate slows even lower than the rate in the first circuit because the total resistance in the RC network has increased.

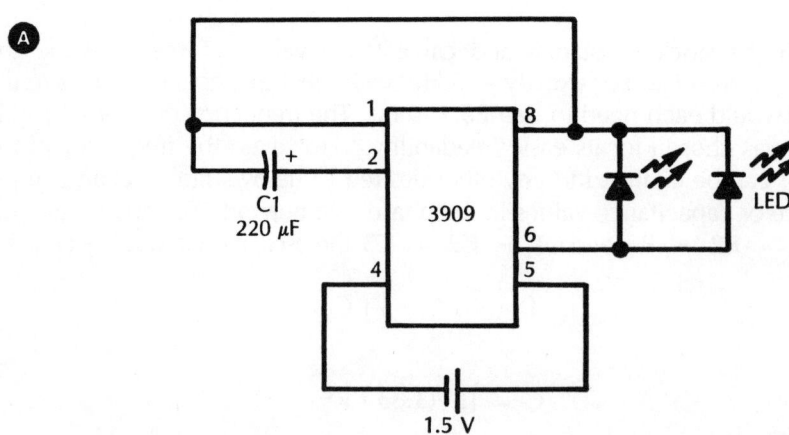

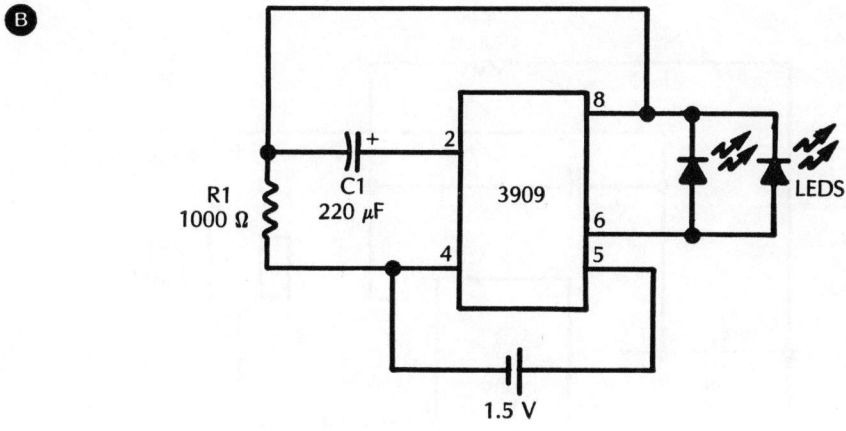

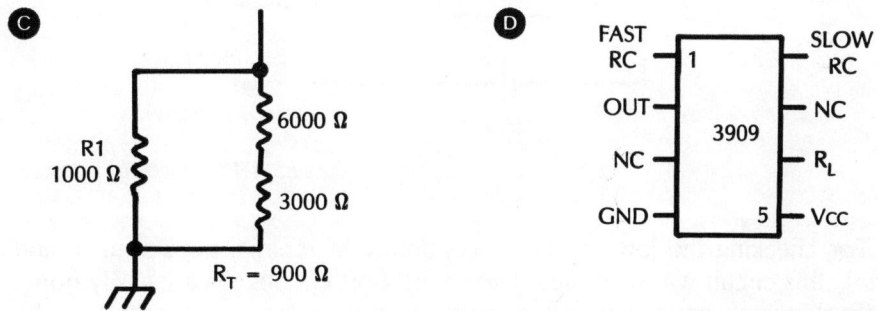

Fig. 4-17. LED blinker.

These blinker circuits can also operate with one LED. With a 3-volt to 6-volt supply, a small incandescent lamp can be flashed with the circuits in Fig. 4-17.

No. 36: Continuity Checker

A very inexpensive, audible continuity checker can be built with the 3909. The circuit in Fig. 4-18 only uses five external components: one resistor, two capacitors, a speaker, and a 1.5-volt battery. By using an AA-size cell, the size of the circuit can be kept small enough to be mounted inside a container that will easily fit in your hand. Any circuit that you check should not have the power applied.

The 8 Ω speaker can be driven directly by the 3909. The test probes can be as simple as two ends of wire, or more elaborate probes can be devised. A high-pitched tone will be heard for a short circuit. The higher the resistance you place between the probes, the deeper sounding the tone. Resistance values above about 2000 Ω will load down the circuit and prevent oscillation. With practice, you might be able to use this circuit as a crude ohmmeter.

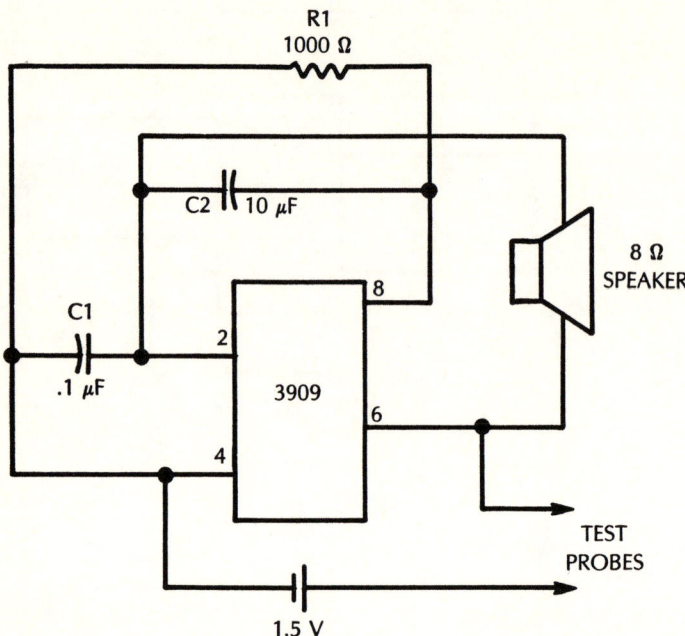

Fig. 4-18. Continuity tester.

For checking the low resistance continuity in motors, transformers, and wiring, this circuit will work fine. Detecting short circuits is very easily done. By checking between two points in rapid succession, differences of 5 Ω can be easily distinguished by the difference in the sound of the tone.

No. 37: AM Radio

The 3909 is a versatile integrated circuit. In Fig. 4-19, it functions as a detector/amplifier in an AM radio. Although the tuning ability is not very good, you should be able to listen to local radio broadcasts. I was only able to receive one station at 1400 kHz. The quality of the audio is about as good as the old crystal sets.

The circuit shown in Fig. 4-19 only uses three capacitors and an antenna coil. The antenna coil is an AM ferrite core salvaged from an old AM radio. I connected about 20 feet of solid copper wire to one end of the antenna coil to improve signal strength. With the 1.5-volt battery, the circuit should operate continuously for about a month. An 8 Ω to 40 Ω speaker of 6 inches or smaller is acceptable.

Using the tuning capacitor (1 to 360 pF) from an AM radio will improve the selectivity and the tuning range of the circuit. The 5 to 150 pF trimmer used in Fig. 4-19 was what I had in my junk box. It works.

Analog ICs 75

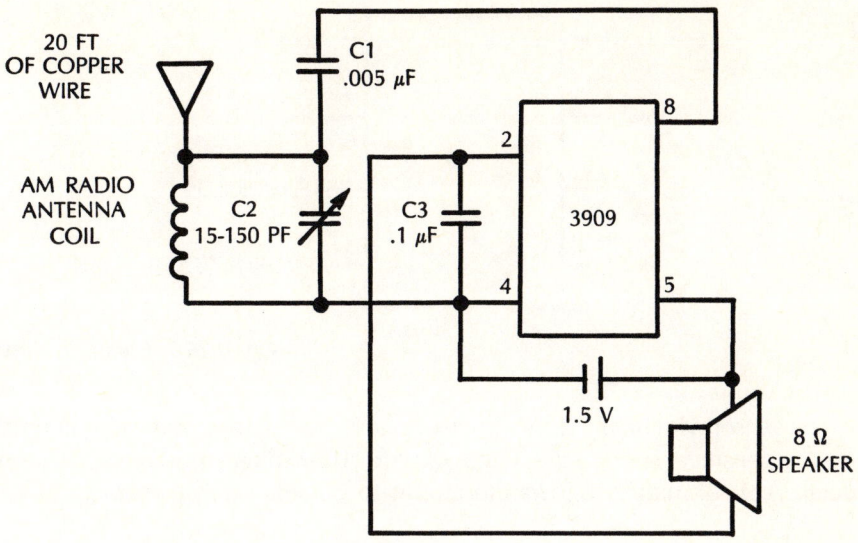

Fig. 4-19. AM radio.

No. 38: Proximity Detector

The circuit in Fig. 4-20 uses the 555 timer IC as a capacitance sensitive switch. A person's body capacitance turns on the switch. When someone gets to within two or three inches of the wire on pin 2, the speaker will begin to click. When the intruder backs away the speaker will again become silent. For instance, if the sensor wire on pin 2 is placed near a door knob, anyone who places their hand on the knob will activate the circuit and set the speaker clicking.

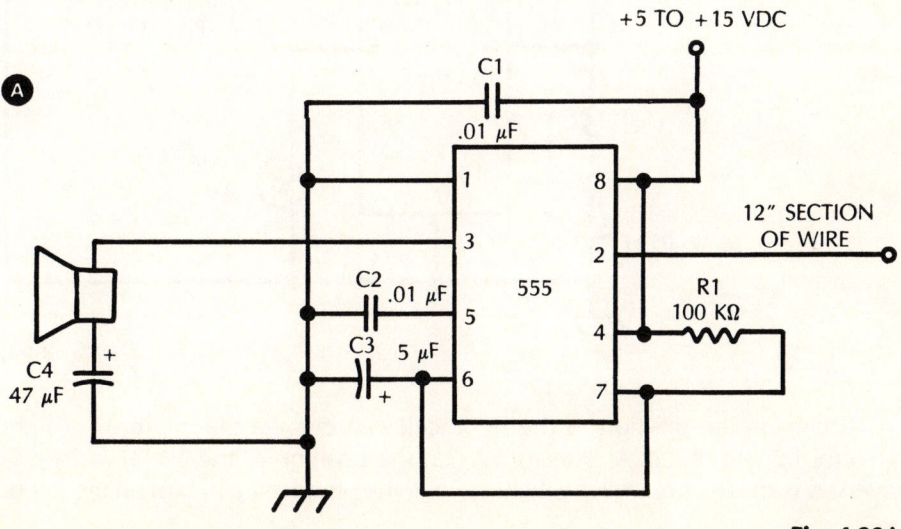

Fig. 4-20A.

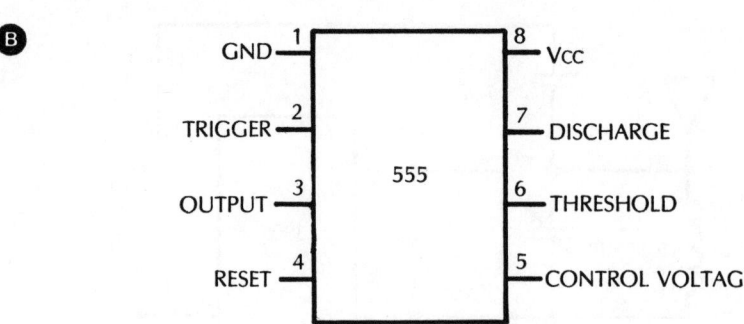

Fig. 4-20B. Proximity tester.

The sensor wire on pin 2 should be about 12 inches long. If the circuit proves to be too sensitive, shorten the length of wire. The voltage requirements are not crucial. A 9-volt battery allows the circuit to be small and portable.

Nos. 39 and 40: Night Light and Electronic Rooster

The cadmium sulfide photocell in Fig. 4-21 provides the light sensitivity needed for the circuit to turn on the lamp in dim to dark conditions. The lamp is a #47. The photocell is available at Radio Shack.

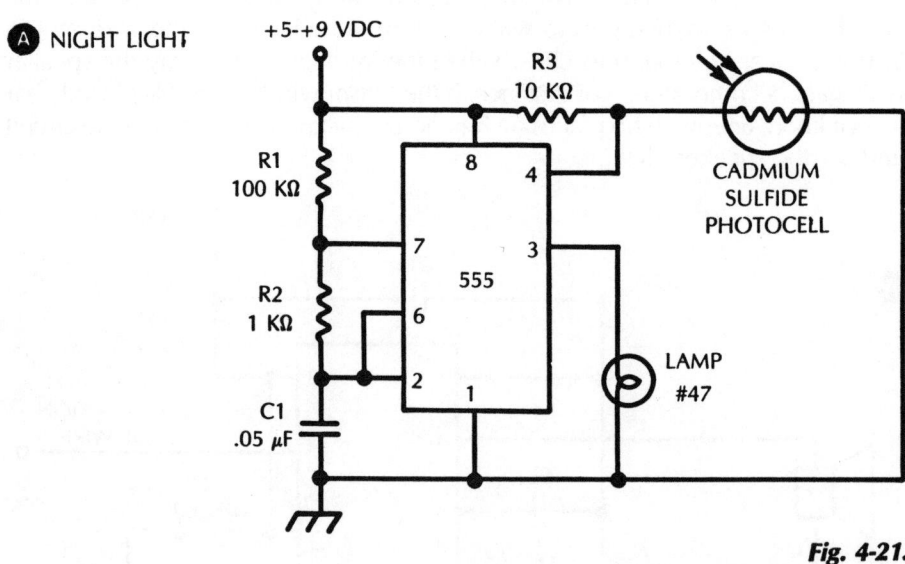

Fig. 4-21.

Changing the position of the photocell can change the circuit to a light detector instead of a dark detector. When the position of the 10 kΩ resistor is reversed with the photocell, the light striking the photocell will turn on the lamp.

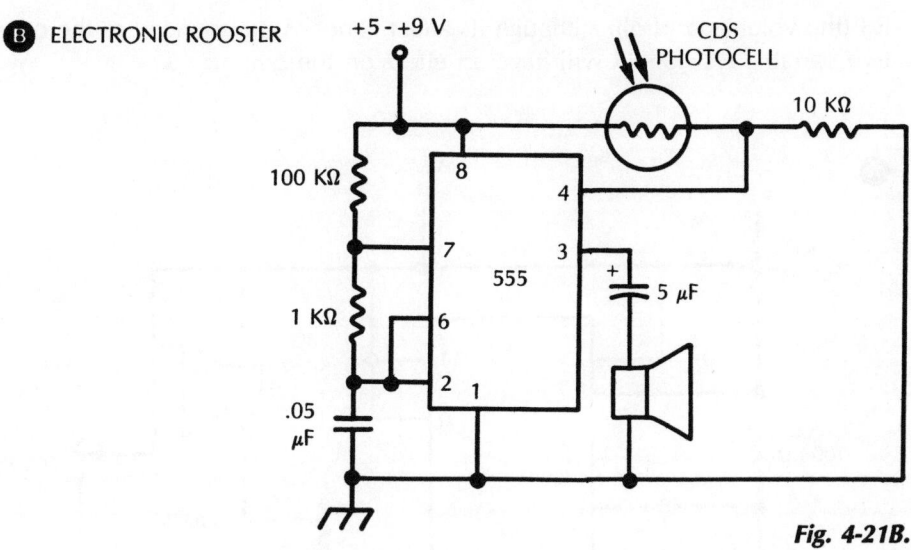

Fig. 4-21B.

This is not a particularly useful circuit. By replacing the lamp with a speaker and placing the photocell in an eastern window, the circuit can become an electronic rooster. Refer to Fig. 4-21B. You will probably want to include an on/off switch in the electronic rooster circuit. Otherwise it will buzz all day long.

No. 41: Electronic Noise Maker

The circuit shown in Fig. 4-22 uses a 556 dual timer. It is a very stable controller that is capable of producing accurate time delays or oscillation. The 556 has two 555 circuits on the same piece of silicon that operate independently of each other. You can operate this IC from +5 to +15 volts. The value of V_{CC}, C1, C2, R1, and C4 determine the time constants for oscillation and timing.

I used a +12 volt supply for the noise maker. Experimenting with different values for C1 and C2 will produce interesting results. Also varying the relative difference between C1 and C2 will give some interesting sounds. The values for C1 and C2 in Fig. 4-22 gave a range of sounds that I thought were very interesting. The circuit is not particular about whether you use electrolytic or ceramic disk capacitors. Use whatever you have lying around in your junk box. Try the following values for C1 and C2.

C1	C2
.05 µF	1 µF
.01 µF	.01 µF
.05 µF	.005 µF

When you have selected a pair of resistors, vary R1 and R4 to obtain an assortment of noises. The number of different sounds will amaze you. Also vary

R3 (the volume control). Although its affect is not as pronounced as the other two variable resistors, it will have an effect on the output.

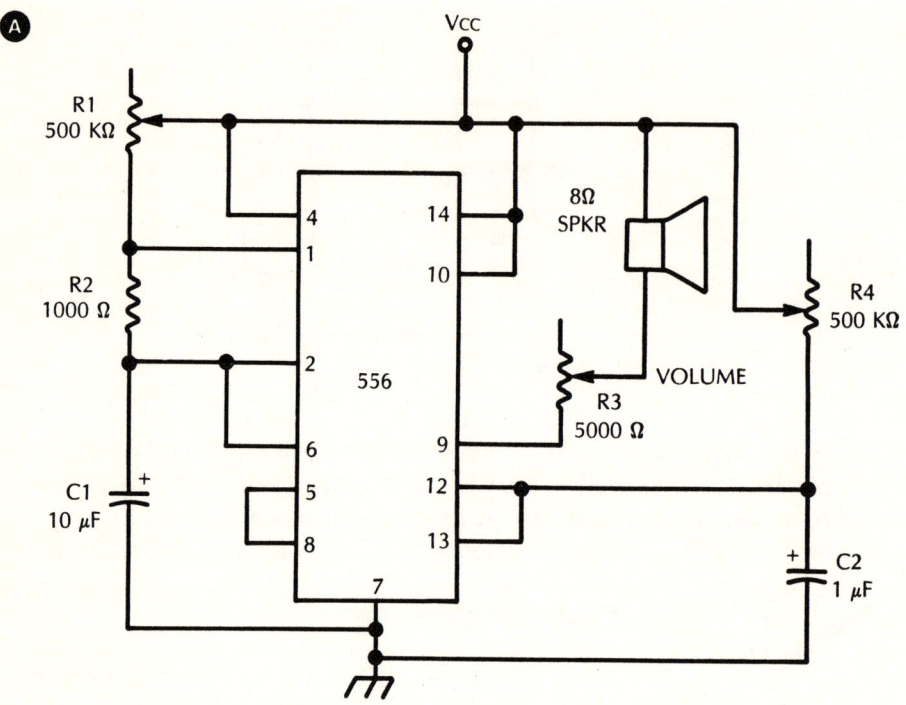

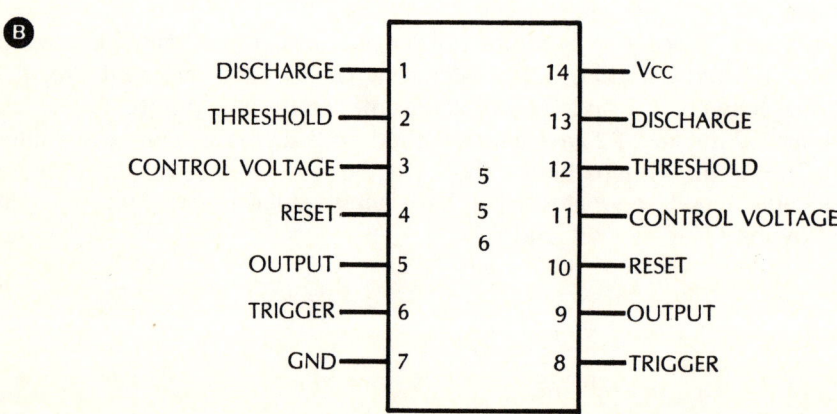

Fig. 4-22. Electronic noise maker.

No. 42: Sound Machine

The circuit shown in Fig. 4-23 exploits the potential of the 94281 Complex Sound Generator. By changing the various switch settings and adjusting the variable resistors, an endless variety of sounds can be created. Sounds ranging from bird chirps to the phasors of the USS *Enterprise* are all possible.

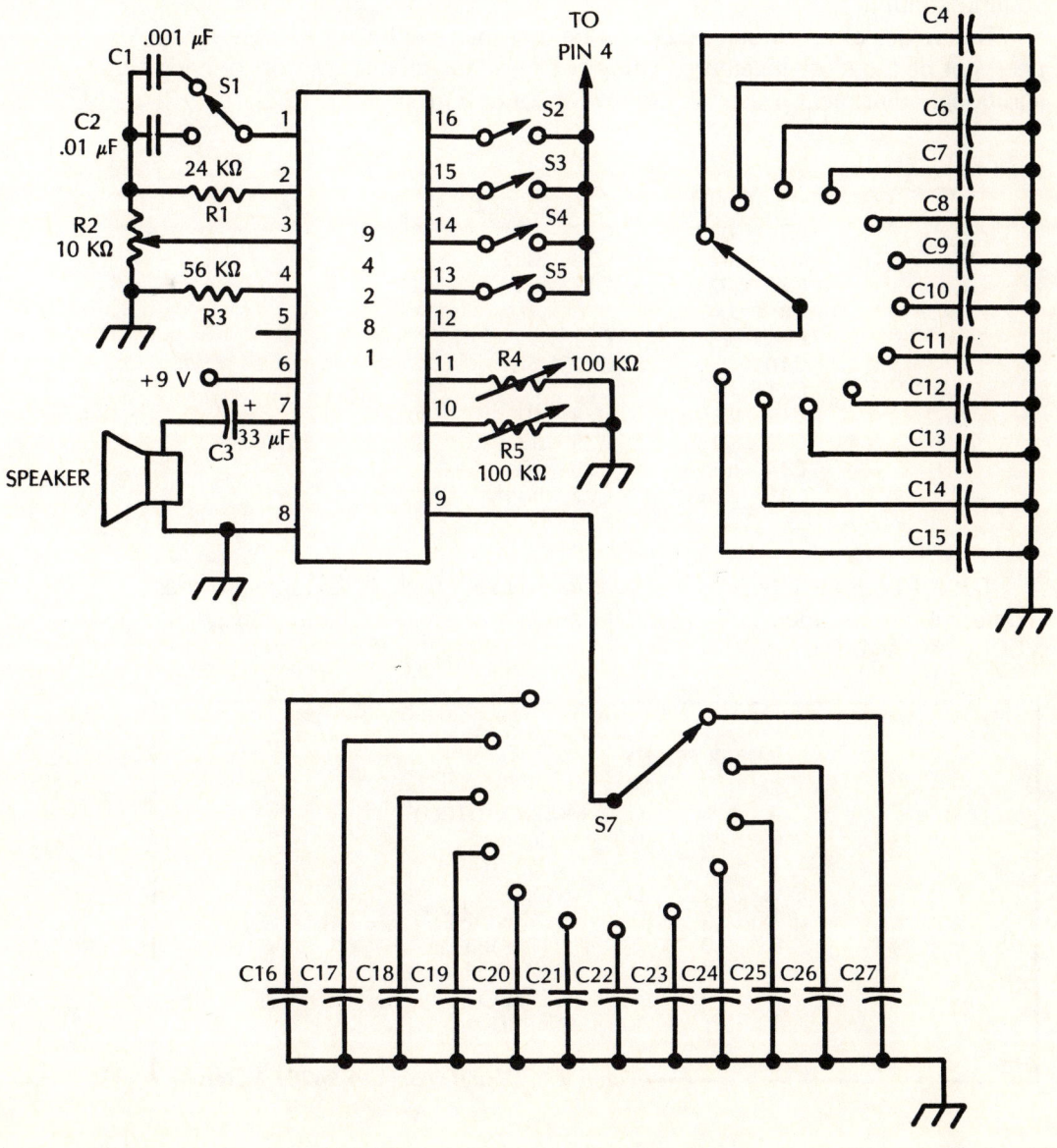

Fig. 4-23.

Switch S1 controls the high and low pass filter. S2 through S5 control the digital logic inputs. These switches rearrange the internal circuitry by changing which internal inputs are fed to the internal mixer and whether or not the VCO is functional. S6 controls the frequency of the Super-Low-Frequency Oscillator by switching in different capacitances. S7 does the same thing as S6 for the Voltage-Controlled Oscillator.

R4 provides finer adjustment of the Super-Low-Frequency Oscillator. R5 accomplishes the same purpose for the Voltage-Controlled Oscillator. R2 is the volume control.

The values of C4 through C27 can be any that you have available. They need not be too close together in value because the variable resistors provide additional adjustment. I used the following values, all in μF:

S6		S7	
C4	.001	C16	.00047
C5	.005	C17	.02
C6	.01	C18	.022
C7	.02	C19	.01
C8	.047	C20	.1
C9	.25	C21	.22
C10	1	C22	.56
C11	10	C23	1
C12	22	C24	10
C13	50	C25	22
C14	none	C26	50
C15	none	C27	none

Table 4-1 lists the input combinations for S2 to S4. Each digital input provides a different mixer output. S5 will inhibit the VCO when low (0) and activate the VCO when high (1).

MIXER SELECT			
S2	S3	S4	
A	B	C	MIXER OUTPUT
0	0	0	VCO
1	0	0	SLF
0	1	0	NOISE
1	1	0	VCO/NOISE
0	0	1	SLF/NOISE
1	0	1	SLF/VCO/NOISE
0	1	1	SLF/VCO
1	1	1	INHIBIT

Table 4-1. Digital Inputs to 94281 Mixer.

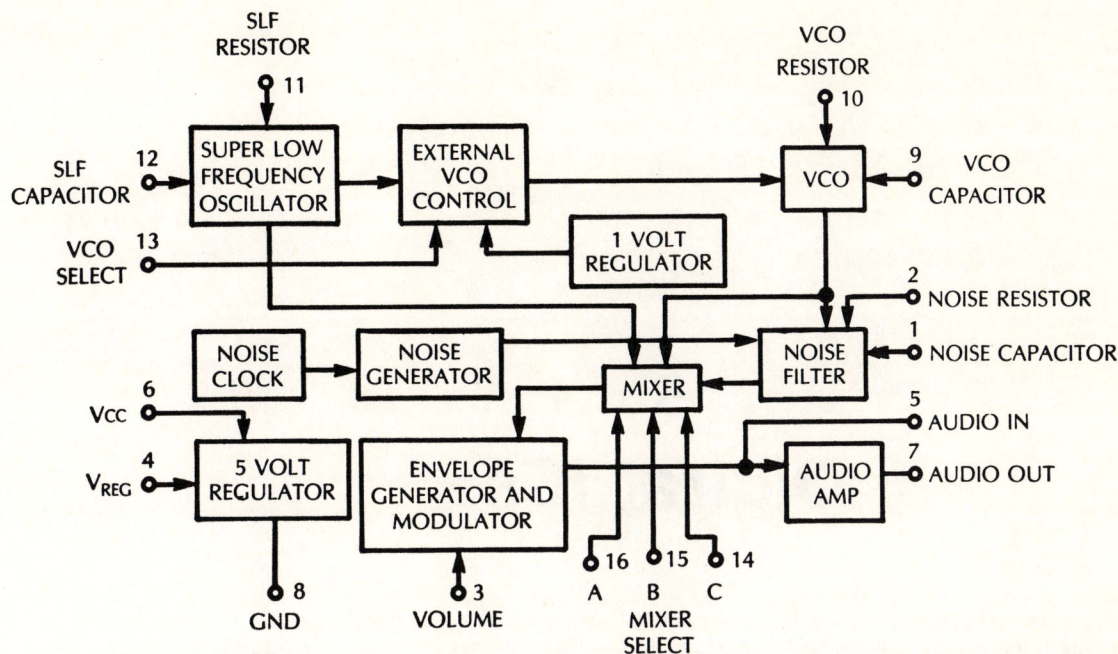

Fig. 4-24. Block diagram of the 94281 Complex Sound Generator.

5
Digital ICs

Digital electronics sets the pace of electronics technology today. Although analog electronics is not dead, it has taken a distant second place. This supremacy of digital electronics is most obvious in the computer revolution. After all, computers are digital machines.

There are two major families of integrated circuits: TTL (transistor-transistor logic) and CMOS (complementary metal oxide semiconductor). Both of these families include digital integrated circuits. The CMOS family also includes some analog ICs and some ICs with both digital and analog circuitry on the same chip.

This chapter will explore the TTL digital ICs. The projects will illustrate their use and readily show how easy they are to use. Chapter 6, which is on CMOS ICs, explores more advanced digital logic functions. Because TTL circuits are functionally the same as CMOS ICs, the basic logic functions will not be repeated. Other than supply voltage differences, some handling precautions, and the rate of power consumption, these two types of integrated circuits are interchangeable and can be used together in the same circuit. Some of the CMOS ICs are pin-for-pin compatible with their TTL counterparts.

THEORY

The transistors in digital ICs are implemented on the silicon in the same manner as those on analog ICs. This is discussed in the preceding chapter. The major difference is that the transistors on a digital chip will either be fully on or fully off. In contrast, the transistors on analog ICs can conduct at an infinite number of levels. Figure 5-1 illustrates signals that might be measured at a transistor on an analog chip and also on a digital chip. Note that Fig. 5-1A shows a continuously varying signal, but Fig. 5-1B shows a signal that can only assume two discrete states (on and off).

Digital ICs 83

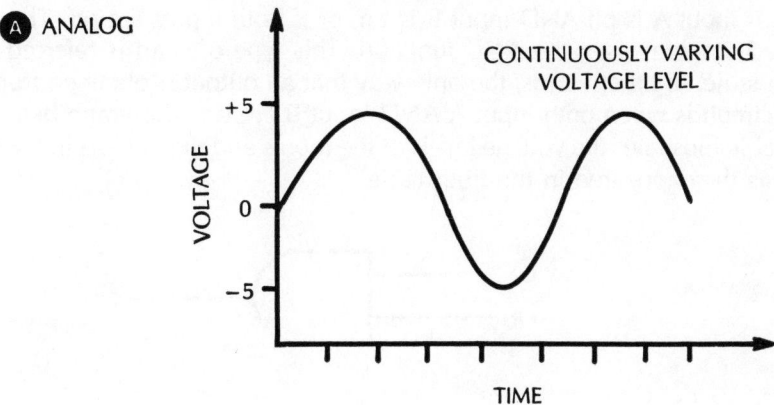

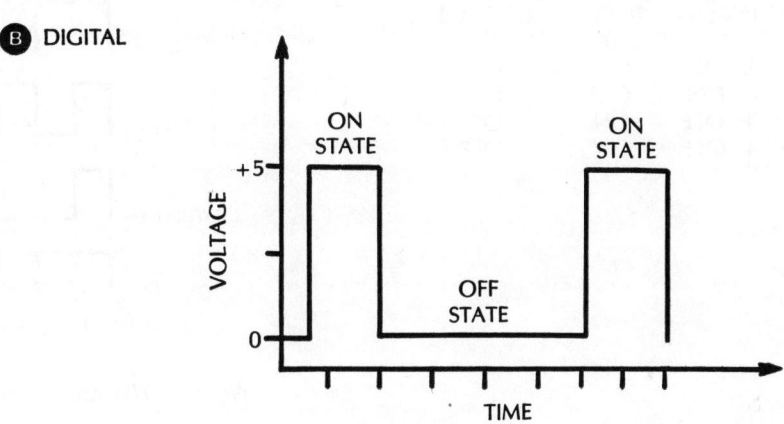

Fig. 5-1. Comparison of an analog and a digital signal.

Digital Logic

This two-state (on/off) condition is the heart of digital logic and digital computers. By controlling the on/off conditions of the transistors on the digital IC, a variety of logic functions can be implemented. There are three basic logic functions: AND, OR, and NOT. Other logic functions can be derived by combining these basic operations.

Figure 5-2 illustrates the AND logic operation. If input A in Fig. 5-2 is on AND input B is on, then output C is on. This is the definition of the AND logic function. If input A is on AND input B is off, then output C is off. The same is true if input A is off AND input B is on, or if both inputs are off. The chart in Fig. 5-2 summarizes the AND function. This type of chart is referred to as a truth table. In other words, the only way that an output is obtained from the AND circuit is when both input A AND input B are on. The graph below the truth table illustrates the voltage levels of the inputs and the outputs in the same order as they appeared in the truth table.

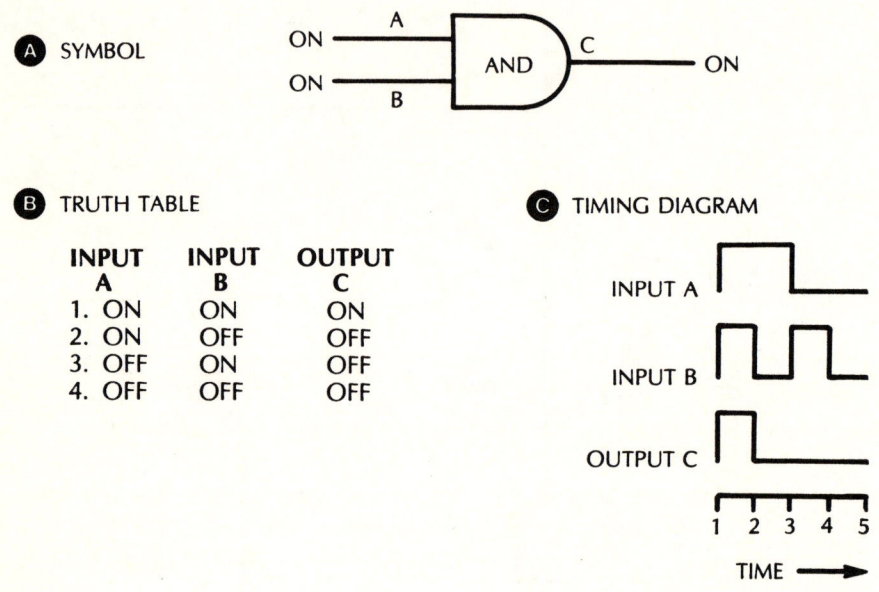

Fig. 5-2. The AND function.

Although the AND concept is quite simple, complex combinations of this and other gates make the digital computer a possibility. Figure 5-3 illustrates two other basic logic functions in the same manner that the AND gate was illustrated. The OR gate shown in Fig. 5-3 will produce an output if input A OR input B is on.

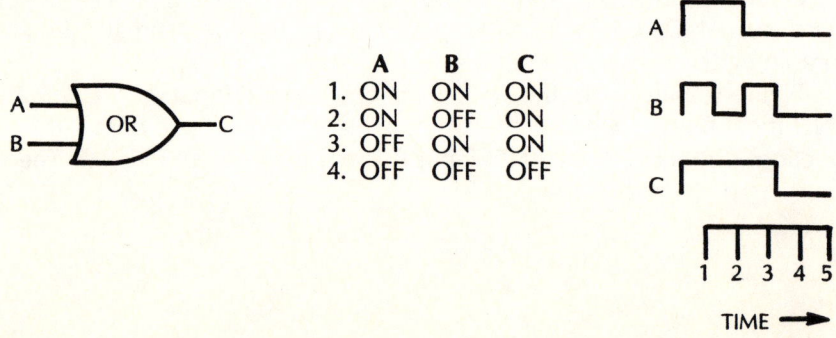

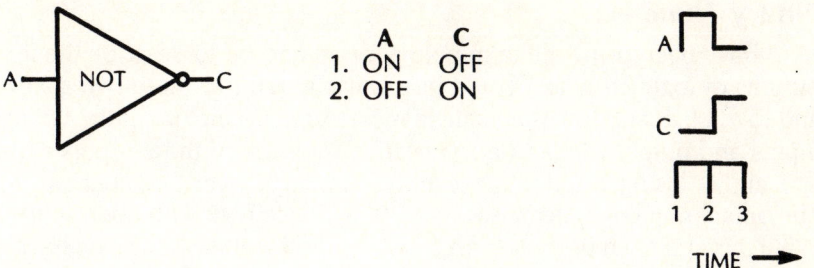

Fig. 5-3. The OR and NOT functions.

The NOT gate in Fig. 5-3 produces an output when input A is off. The NOT gate is a signal inverter; output C is always the opposite of input A. When the NOT gate is combined with the AND gate or the OR gate, NAND and NOR gates result respectively. Figure 5-4 illustrates these two additional gates.

Digital circuits can be used to perform logic operations of AND, OR, and NOT. By combining these fundamental operations, circuits can be built to simulate decision-making. The first project in this chapter illustrates the use of digital circuits performing logic operations.

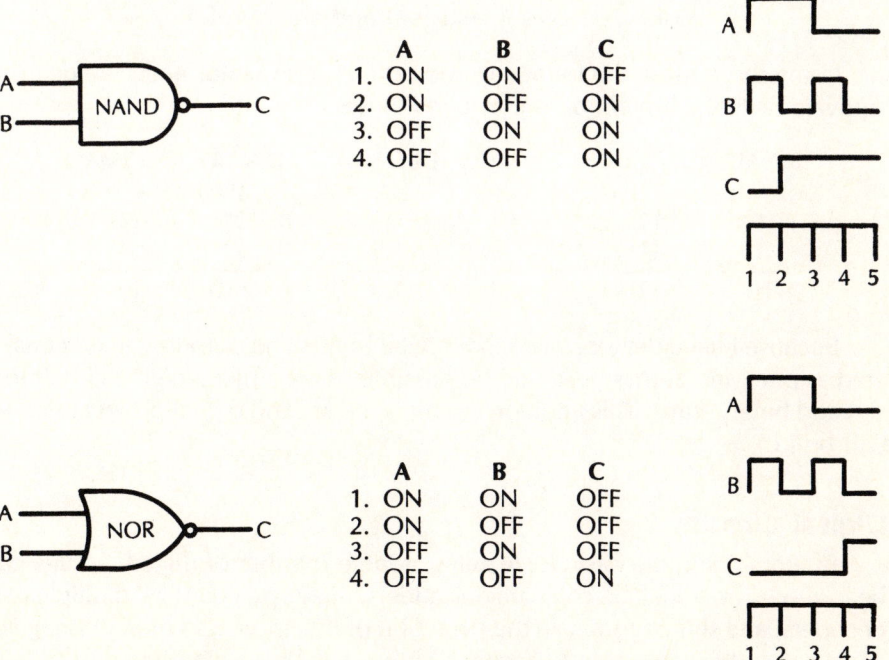

Fig. 5-4. The NAND and NOR functions.

Binary Numbers

I have been using the terminology of on and off to describe the inputs and outputs of logic circuits. I could just as easily refer to these two states as high and low. High and low is actually a more accurate picture of the voltage at the inputs and outputs. I could also describe the state of these inputs and outputs as 1 and 0. Using 1 and 0 has several advantages over the other two methods. The binary number system uses 1 and 0 as the only two numbers in the system. What this approach ultimately accomplishes is the ability to use digital electronic circuits to perform binary math.

A quick review of the binary number system and the decimal equivalents might be necessary. The following table compares the decimal numbers with their binary counterpart.

BINARY	DECIMAL	BINARY	DECIMAL	BINARY	DECIMAL
0	0				
1	1	100	4	111	7
10	2	101	5	1000	8
11	3	110	6	1001	9

To add binary numbers, the following rules are observed.

RULE 1. 0 plus 0 equals 0
RULE 2. 0 plus 1 equals 1
RULE 3. 1 plus 0 equals 1
RULE 4. 1 plus 1 equals 0 and a carry of 1

Using these rules, the following additions can be performed. The decimal equivalents to each addition are in parentheses.

BINARY	(DECIMAL)	BINARY	(DECIMAL)	BINARY	(DECIMAL)
				1 carry	
1	(1)	01	(1)	11	(3)
1	(1)	10	(2)	11	(3)
10	(2)	11	(3)	110	(6)

Because I have demonstrated that digital inputs and outputs can be considered as ones and zeroes, it should be possible to construct a digital circuit that can add binary sums. This addition circuit is the second digital project that you will build.

Digital Circuits

By combining the basic logic gates, a large number of digital circuits can be formed. These include comparators, adders, flip-flops, counters, multiplexers, decoders, and shift registers. In the evolution of ICs, more and more basic gates were put on the same chip to accomplish more sophisticated functions. Now, instead of having to build up a counter from several NAND gates, the entire counter is integrated on one chip. This progression has continued to the point where 100,000 transistors can be placed on the same chip to create a computer.

But even a sophisticated microprocessor can be considered in terms of the basic logic gates. Because of the great number of gates, however, the logic gate level is too complex for all but the hardiest individuals. Instead, the microprocessor may be more successfully understood by grouping the basic logic gates together into their specific functions like registers, adders, shift registers, and counters. For some applications, only the inputs and outputs of the microprocessor may have to be considered.

This macroscopic approach points up the beauty of digital circuits. They can be treated as black boxes. Consideration to only the inputs and the outputs might be all that is required for most circuits. The more sophisticated ICs with many different functions integrated on the same chip become building blocks to even more complex circuits. The projects in this chapter and the next chapter illustrate this principle.

No. 43: Magnitude Comparator

Figure 5-5 demonstrates a digital circuit that can determine which of two inputs is numerically greater and can indicate when both inputs are equal. The circuit uses four AND gates and two NOT gates. LEDs are used to indicate the result of the comparison. The circuit does not actually decide anything. The trick is in the wiring. With two inputs to compare, the circuit needs to deal with the following input combinations.

A	B
0	0
1	0
0	1
1	1

These are the only combinations of inputs that can appear on input A and input B. The circuit ANDs input A and input B together in four different ways. AND gets 1 compares the input on A with the inverse of the input on B. You can see that the only way that AND gate 1 can have a high output is when both of its inputs are high. This means that if input A is 1 and input B is 0, then the LED indicating that A>B will glow. The outputs of all the other AND gates will be low. Check the circuit to be sure. The following chart indicates which LED will glow for each input combinations.

A	B	A>B	A=B	A<B
0	0	no	yes	no
1	0	yes	no	no
0	1	no	no	yes
1	1	no	yes	no

Try each combination to ensure that the proper LED glows. This circuit provides a simple glimpse into the use of digital circuits in logic functions. This circuit is a simplified explanation of how two values in a computer program would be compared in an IF THEN statement.

IF A=B THEN 1000

In this statement the output of a comparator circuit is looked at to see if the statement A=B is true or not. If the output is 1 or high, then the statement is true and the THEN portion is executed. The only major difference between the computer logic circuit and the one in Fig. 5-5 is the number of inputs that it compares.

The circuit in Fig. 5-5 only compares one bit (A compared to B). The comparator in a CPU would probably compare at least eight bits against each other. Figure 5-5 can be expanded to compare more bits by increasing the number of inputs and the number of AND and NOT gates. The tangle of connections quickly gets out of hand to draw such a circuit. By using an IC that has the comparator already implemented on the chip all of this tangle can be eliminated.

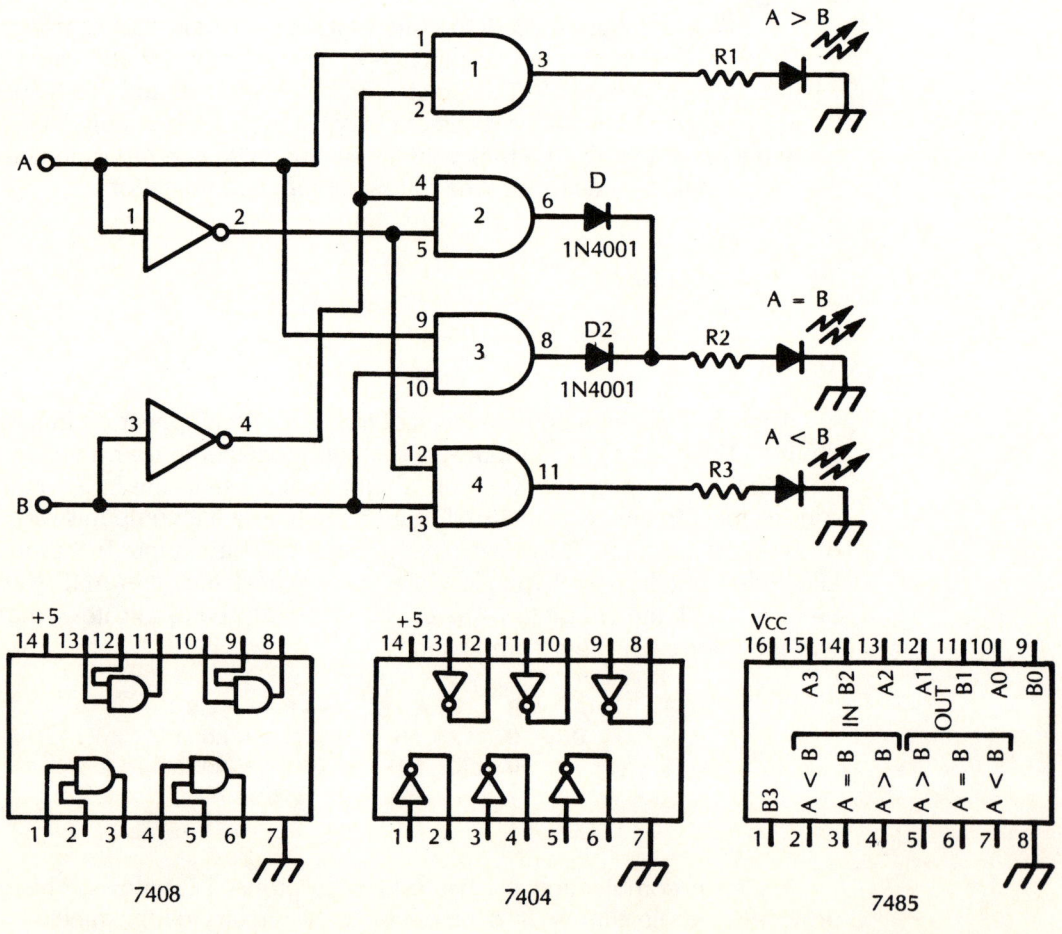

Fig. 5-5. Simple comparator.

The 7485 4-bit comparator is just such a chip. This IC can compare two sets of 4 bits of information (called a nibble) and indicate which is larger and whether or not the inputs are equal. And it can do all of this with less external wiring than that in Fig. 5-5.

The pinouts for the 7408 AND gate, the 7404 NOT gate, and the 7485 comparator are shown at the bottom of Fig. 5-5. The pin numbers on the circuit diagram correspond to those on the pinouts for each chip. Both chips use a 5-volt source. D1 and D2 provide isolation for the outputs of gate 2 and gate 3. The three resistors can be 330 Ω to 1000 Ω. These resistors provide current limiting for the LEDs.

No. 44: Half Adder

Digital circuits can also be used to perform arithmetic functions. The circuit shown in Fig. 5-6 performs binary addition on the two inputs. To do this, five NAND gates are needed (NOT AND gates). You can recognize the NOT circle on the end of the AND gate that indicates the NAND function. Just as in the last project, the secret is in the wiring. The truth table below the circuit shows all the possible outcomes of adding two 1-bit binary numbers. All that is required is building a circuit whose outputs meet the truth table.

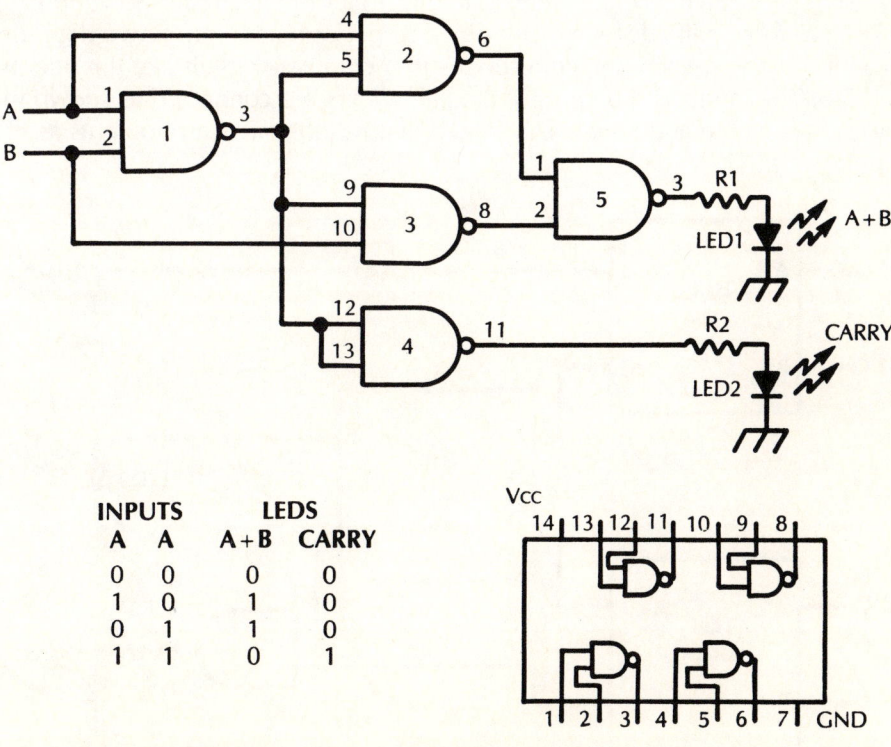

INPUTS		LEDS	
A	A	A+B	CARRY
0	0	0	0
1	0	1	0
0	1	1	0
1	1	0	1

7400

Fig. 5-6.

Consider the following addition.

$$\begin{array}{r} 1^{\text{carry}} \\ 1 \\ +1 \\ \hline 10 \end{array}$$

This addition is the same as the last line in the truth table in Fig. 5-6. Follow the inputs in the circuit through to the outputs. When input A and input B are both 1, the output from gate 1 is 0. Check this result with Fig. 5-4 if necessary. Because the inputs to gate 4 are tied together, they are also both 0, and the output of gate 4 is 1. A 1 output will cause the LED on the carry line to glow. Gates two and three both have a 1 and a 0 input which produces a 1 output from each gate. These two outputs give both of gate five's inputs a 1, resulting in a 0 output from gate five. A 0 output keeps the A+B LED dark. The LEDs, then, indicate a binary 10 result of the addition, or 2 decimal, which is the correct answer. Try the other possible sums to verify that the circuit works.

No. 45: Full Adder

Although there are only four combinations that can be added with this circuit, it can be expanded to include more inputs by using combinations of half-adder circuits. Refer to the circuit shown in Fig. 5-7. Here three half adders are combined to perform addition of two-digit, or two-bit, numbers. Each box corresponds to a complete circuit like the one in Fig. 5-6. The inputs and outputs of the half adders are connected as shown in Fig. 5-7. Note the additional NAND gate that has the two carry outputs as its inputs.

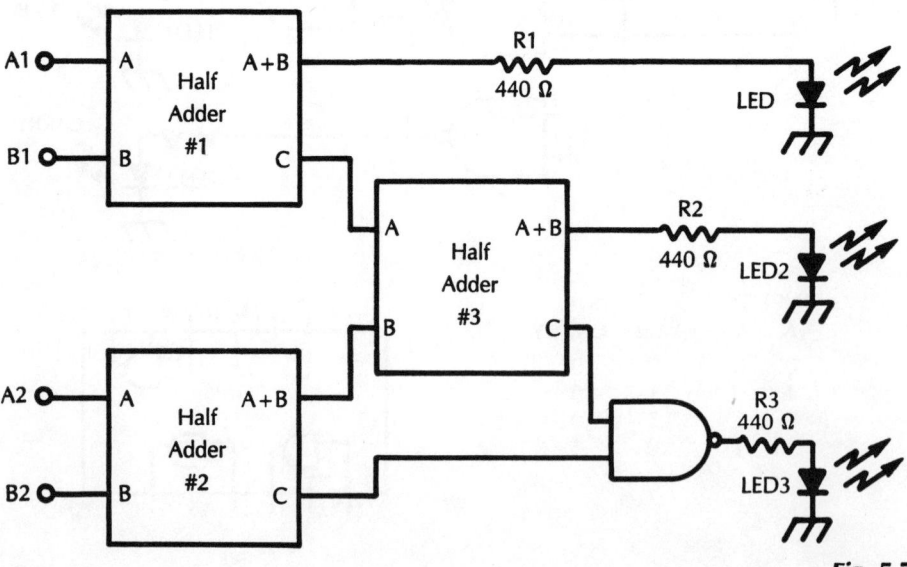

Fig. 5-7.

Table 5-1 lists all the possible sums for a two-bit addition. The ones and zeroes on the inputs and outputs in Fig. 5-7 correspond to the last line of the truth table in Table 5-7. The addition looks like this.

```
   INPUTS      BINARY    DECIMAL
                1 carry
    A2 A1        11         3
   +B2 B1       +11        +3
                ---        --
                110         6
```

A2	A1	B2	B1	LED3	LED2	LED1	DECIMAL SUM
0	0	0	0	0	0	0	0
0	1	0	0	0	0	1	1
1	0	0	0	0	1	0	2
1	1	0	0	0	1	1	3
0	0	0	1	0	0	0	1
0	1	0	1	0	1	0	2
1	0	0	1	0	1	1	3
1	1	0	1	1	0	0	4
0	0	1	0	0	1	0	2
0	1	1	0	0	1	1	3
1	1	1	0	1	0	1	5
0	0	1	1	0	1	1	3
0	1	1	1	1	0	0	4
1	0	1	1	1	0	1	5
1	1	1	1	1	1	0	6

Table 5-1. Truth Table for the Full Adder.

When inputs A1 and B1 are both 1, you know from the truth table in Fig. 5-6 that the A+B output is a 0 and the carry is a 1. Figure 5-7 shows this result. When the A2 and B2 inputs are both 1, again you know that this input condition to a half adder gives a 0 output at A+B and a 1 output on the carry. A 0 on the A+B line keeps LED1 turned off.

Half adder #3 gets a 1 input from half adder #2's carry output. It also gets a 0 input from the A+B output of half adder #2. A 1 and 0 input to a half adder produces a 1 output on A+B and a 0 output on the carry. The 1 output on A+B turns LED2 on. The 0 output on the carry of half adder #3 is NANDed to the 1 output on half adder #2. A 0 and a 1 input to a NAND gate produces a 1 output and turns on LED3. Compare the other inputs and outputs in Table 5-1 to the functioning of the circuit in Fig. 5-7.

The addition circuitry in a microprocessor uses the same principles as this simplified binary addition circuit. The addition circuitry is located in the Arithmetic Logic Unit in a microprocessor. A microprocessor can add much larger numbers because it has more half adders strung together.

But just as the circuits in Fig. 5-6 and 5-7 had a limit to the size of the number they could add, so does the microprocessor. Special software routines can increase the size of the number-handling capability even further.

No. 46: Binary Counter

The circuit shown in Fig. 5-8 illustrates how digital circuits can be used to count. The 7476 dual flip-flop contains 2 separate flip-flops. By using two 7476 chips, you have four flip-flops at your disposal. By connecting the Q output from one stage to the clock input of the next stage, a binary counter is constructed with a maximum count of 16 (0 to 15). Note the truth table for the counter in Fig. 5-8.

The 7476 is a versatile chip, and this circuit uses only one of its features. When the J and K inputs are held high and the preset (P) and clear (C) inputs are also held high, the Q output will toggle back and forth between 1 and 0 with every pulse that appears on the clock input. In essence, each stage of the counter only counts from 0 to 1 and then starts over again. Another way to describe the action is to say that each stage divides the clock frequency by two. You can see this division in the truth table of Fig. 5-8. The time that each LED spends in the on state is increased by a factor of two.

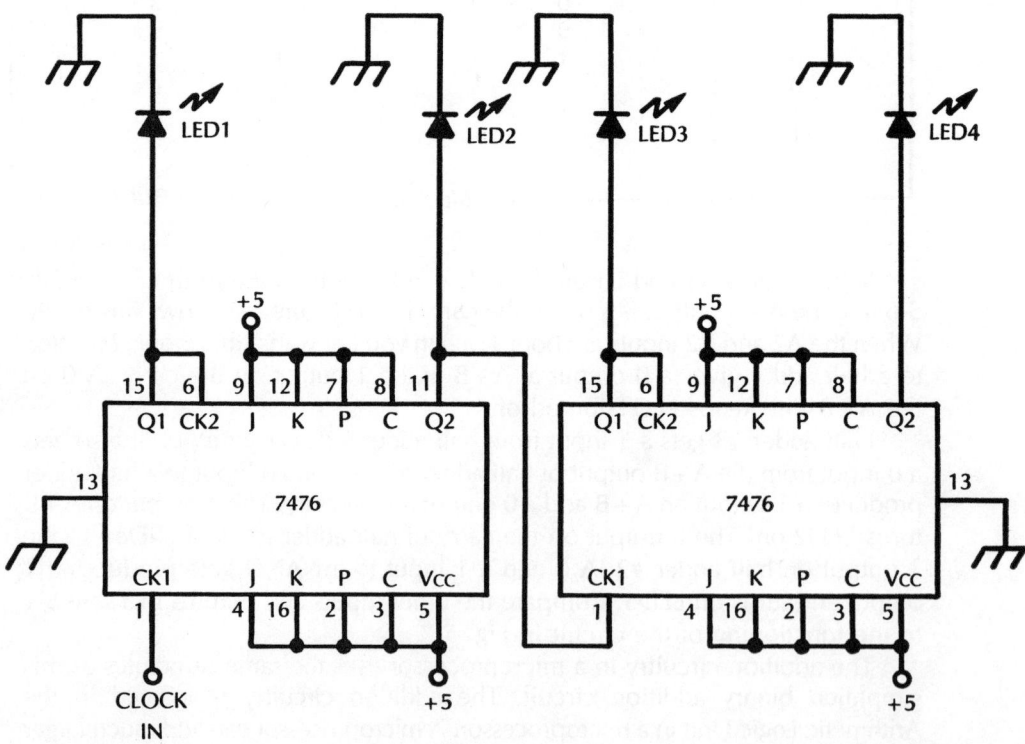

Fig. 5-8.

CLOCK COUNT	LED4	LED3	LED2	LED2
0	0	0	0	0
1	0	0	0	1
2	0	0	1	0
3	0	0	1	1
4	0	1	0	0
5	0	1	0	1
6	0	1	1	0
7	0	1	1	1
8	1	0	0	0
9	1	0	0	1
10	1	0	1	0
11	1	0	1	1
12	1	1	0	0
13	1	1	0	1
14	1	1	1	0
15	1	1	1	1
16	0	0	0	0

Truth Table for the Binary Counter.

The principles of binary counting shown by the circuit in Fig. 5-8 are used in computer circuits to count. The program counter in the CPU of a computer is doing exactly this. After executing each instruction, the program counter increments by one, and the next instruction is taken from the memory location indicated by the counter. In this fashion, program statements are executed one after another in sequential order. Program jumps and loops require some additional circuitry, but the basic concepts are the same.

Note that the circuit in Fig. 5-8 does not have any external components other than the LEDs. This is the power and strength of integrated circuit design. All the necessary components are continued inside the IC. There are a lot of interconnections to make between the IC pins, however. But even this inconvenience can be eliminated by going to the next higher level of integration. An LSI (large-scale integration) binary counter has all four flip-flops and all the interconnections already made on the chip. Other than the connections to the LEDs, all of Fig. 5-8 is contained within a single chip. The next project illustrates the use of an LSI binary counter.

No. 47: LSI Counter

The circuit in Fig. 5-9 accomplishes the same function as the one in Fig. 5-8. It is not difficult to realize that it is easier to construct this second counter. Most of the design work has already been done for you by the IC designer.

The truth table for Fig. 5-9 is identical to the one for Fig. 5-8. Every clock pulse on pin 14 of the 7493 binary counter shows up as a binary count on the four LEDs. The 7493 can count from 0 to 15. Pins 2 and 3 are resets.

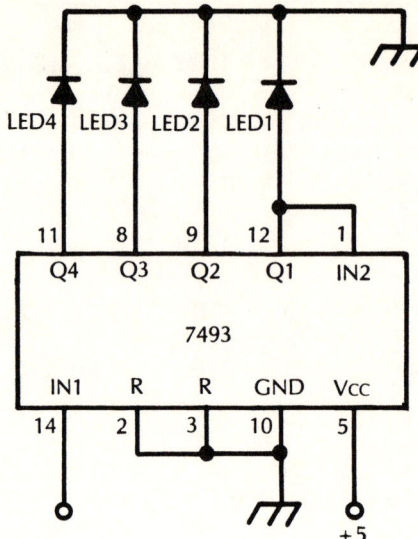

Fig. 5-9. LSI counter.

If a high is placed on either one or both of these pins, the counter will reset to binary zero. It will ignore all clock pulses until both of the resets are low.

No. 48: Binary Logic Probe

When working with digital circuits, it is often necessary to determine the logic state of a particular pin on an IC. The circuit in Fig. 5-10 will give a visual indication of a 1 or a 0 on the 7-segment display. If the probe is placed at a location with a changing state, the 0 and 1 indication will alternately flash on and off. If the frequency of the input is greater than a few hertz, the 7-segment display will indicate a p. Although this logic probe is rather simple, it can be used to do a variety of testing on digital circuits.

Half of a 7404 hex inverter IC is used to distinguish between a 1 or a 0 input on the probe. Before a signal is placed on the probe, the display will indicate a 1. If a low is placed on the probe, inverter 3 will invert the signal and place a low on pin 1, 13, and 11. This will illuminate ¾ of the 0. The back side of the 0 is always lit up because it is always connected to +5 volts (pin 2). If a high appears at the probe, inverter 1 and 2 place a high on pin 7, and the bottom segment of the 1 is lit up. If a pulse appears on the input, the display will indicate a P, for pulse.

The 7-segment display is a common-anode device (MAN 72A). This means that a high, or 1, will illuminate the segments. If a display other than the MAN 72 is used, simply connect the inverters to the pins that light up the proper segments. The resistors shown in Fig. 5-10 are essential. Most 7-segment displays operate at 1.5 volts. R1, R2, and R3 drop the 5-volt supply voltage to a safe level. If you mistakenly connect 5 volts across one of the segments, you will destroy that segment.

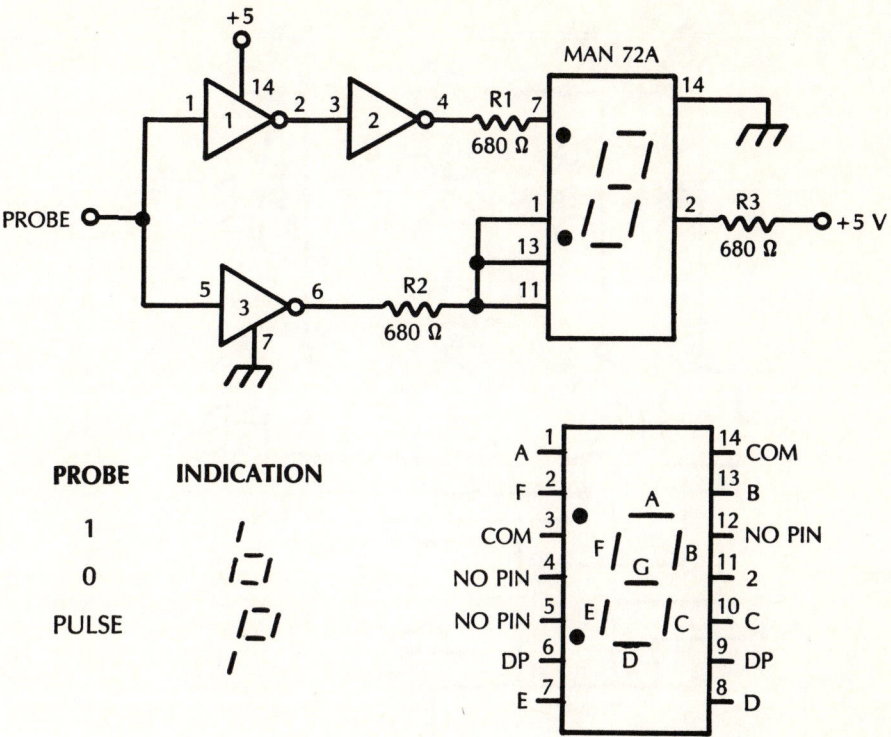

Fig. 5-10. Logic probe.

No. 49: Electronic Dice

By arranging six LEDs in the shape of a die, this project can simulate the toss of a die. When a normally open switch is pushed, a toss is made, and a particular arrangement of LEDs will remain lit. The number lighted indicates the number on the die for that throw. When the button is released, all the LEDs are lighted, indicating that the circuit is working.

Figure 5-11 shows the circuit for electronic dice. NAND gates A, B, and C comprise a 1 MHz oscillator. The 7493 counter has already been used in several previous projects. It is connected in a slightly different way this time. Note that pin 1 is used as the input and that pin 12 is not used as an output. Only three of the four flip-flops inside the 7493 are used. Its outputs are decoded by the NAND gate and the four NOR gates. These gates then drive the six LEDs. Note that a toss of 0 or 7 is not possible. Also note that 1 and 2 occur twice in the count sequence.

Two of these circuits can be used to replace the dice in most games. It is not necesssary to have a second oscillator for the second counting and decoding circuit, however. A single oscillator can be used to drive two counters. Again, note the small number of external components needed to implement this circuit.

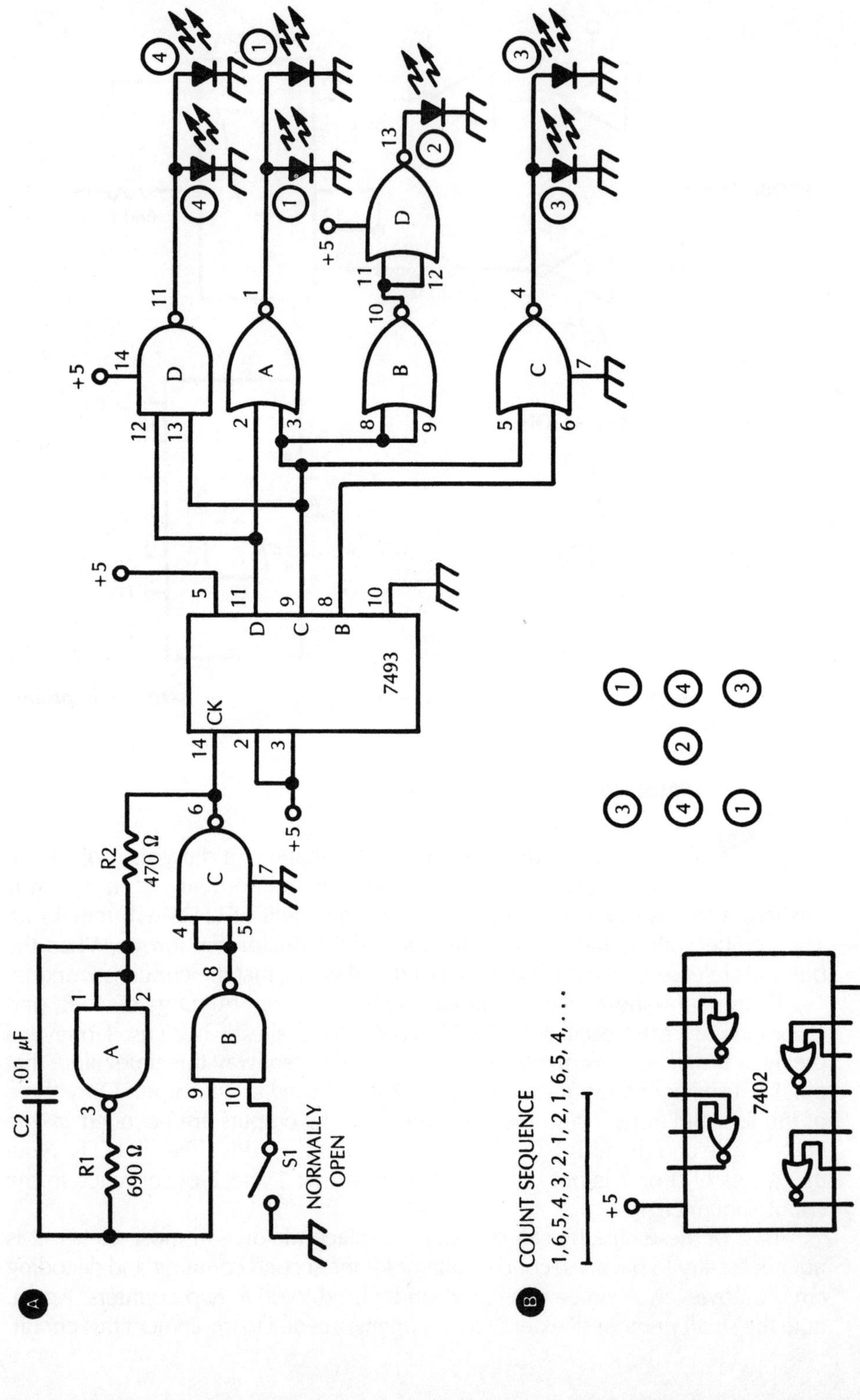

Fig. 5-11. Digital dice.

The pinout for the 7493 is shown in Fig. 5-9. The pinout for the NAND gate is shown in Fig. 5-6. The 7402 pinout is shown at the bottom of Fig. 5-11.

No. 50: Circle Chase

This project illustrates a shift register. By arranging a series of LEDs in a circle, the effect of circular motion can be achieved. Alternately, by placing the LEDs in a line, a pulsing, linear motion is achieved. The circuit shown in Fig. 5-12 uses four 7476 flip-flops. This arrangement will control eight LEDs. More stages can be added by connecting Q outputs to J inputs and Q NOT outputs to K inputs. The clock and clear connections for additional stages are connnected in the same manner as those in Fig. 5-12. Note that the very first stage has its clear pin connected to +5 volts instead of the switch. This connection provides the first pulse that is passed down the chain.

Each flip-flop passes its output on to the next flip-flop every time S2 is pressed. Because the reset switch is wired to preset the circuit with only one lit LED, only one flip-flop (FF1) has a high to pass to the next flip-flop. All the others pass lows. In other words, only one high is passed around the ring. If you do not reset the circuit before pressing S2, every other LED will be lighted. Then when you press S2, four highs will be passed around the ring. Using the reset allows a clearer illustration of the shifting action.

Although this circuit merely flashes LEDs, it illustrates the principle of a shift register—which is another very important digital circuit. In a computer, shift registers are used to perform shift and rotate functions. Practical uses of the shift function are multiplication and exponentiation.

Consider the following examples:

MULTIPLICATION	BIT SHIFTING
2	BINARY 0010 0100
×2	↲
4	DECIMAL 2 4
	The 1 in bit 2 is shifted one place to the right. The result equals decimal four

Simply by shifting the binary one in position two (0010) to position three (0100), the binary representation of decimal 2 is changed to the binary representation of decimal 4. In other words, binary multiplication by two can be performed simply by shifting bits to the left. Consider this next example:

EXPONENTIATION	BIT SHIFTING
$2^3 = 8$	BINARY 0010 1000
	↲
	DECIMAL 2 8
	The 1 in bit 2 is shifted two places to the right. The result equals decimal eight.

98 Beyond the Transistor: 133 Electronics Projects

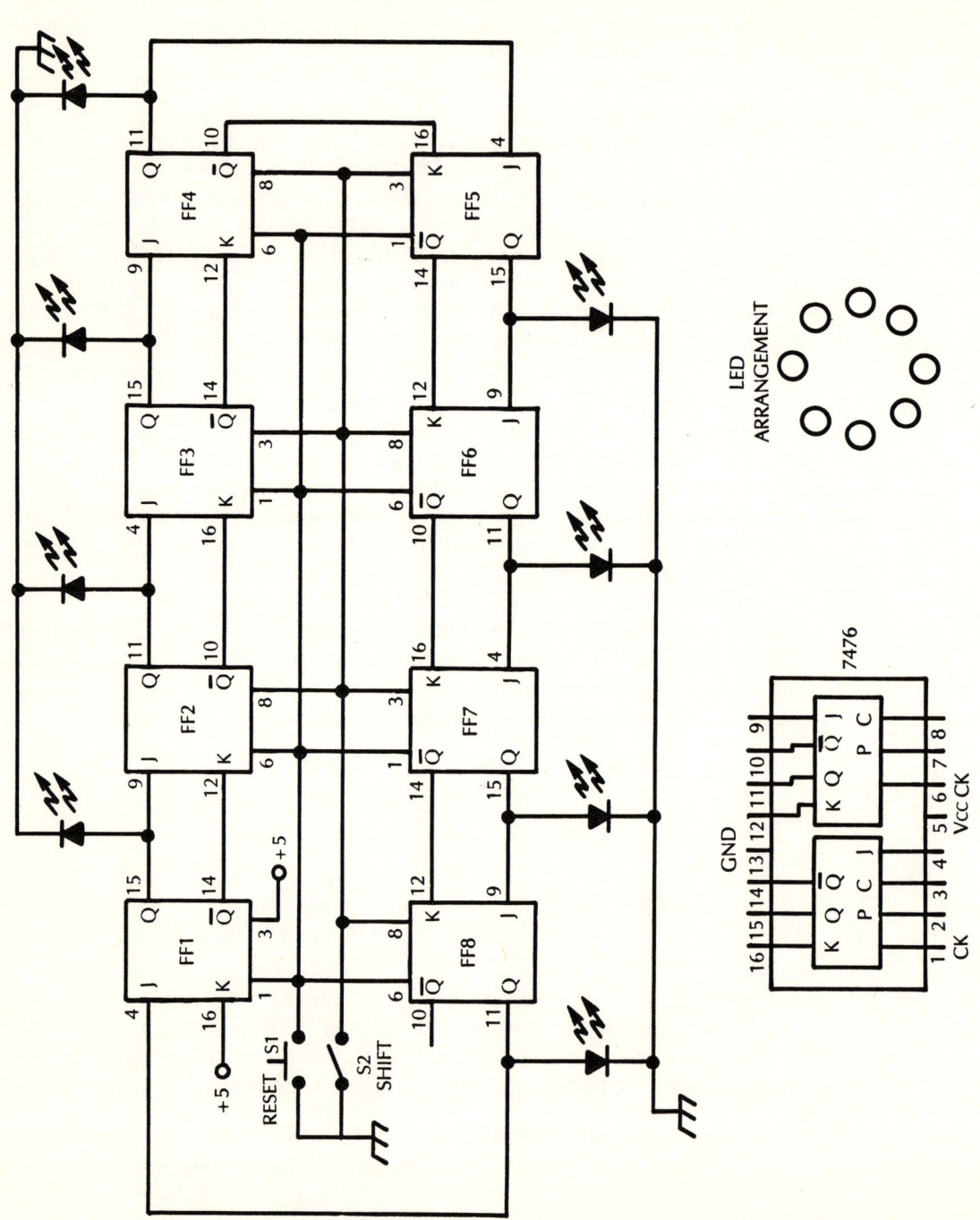

Fig. 5-12. Circle chase.

Again, just by simple bit manipulation, a number can be raised to a power of two. Also, note that by shifting a bit one place in the opposite direction is equivalent to taking the square root of a number.

No. 51: Melody Maker

Although Fig. 5-13 looks a little complicated, the circuit itself is not. One-half of the 556 dual-timer is set up as a low-frequency oscillator. R1 and C1 control the frequency of this oscillator. The other half is set up as an audio oscillator. R12 controls the base frequency that the oscillator will produce. The low-frequency output from the first half of the 556 steps the 7493 counter through its count sequence. The four binary outputs of the 7493 are decoded by the 7448 decoder/driver into 9 outputs. (The 7448

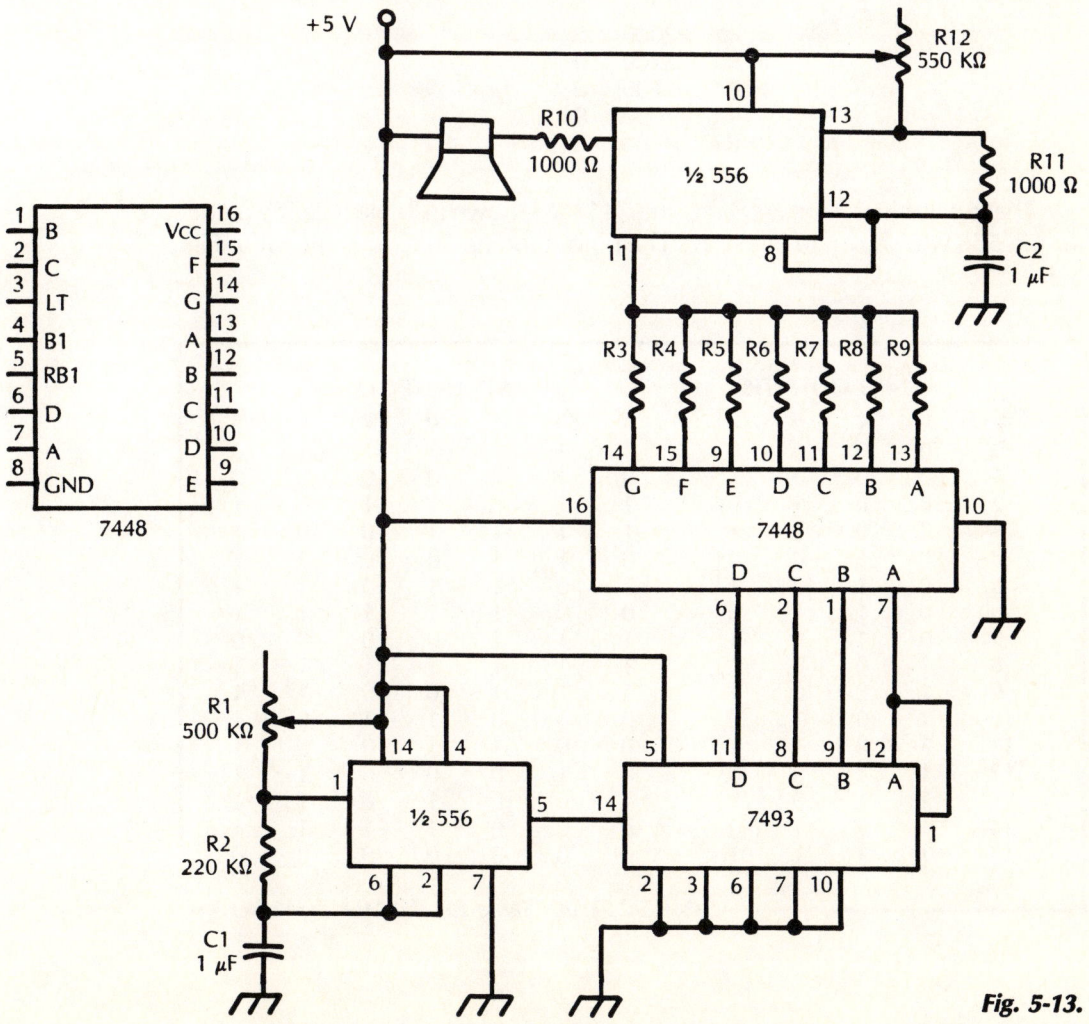

Fig. 5-13.

is not normally used in this manner.) Its primary use is to drive a seven-segment display. It is used as a display driver in Chapter 6 and Chapter 10.) Each discrete count of the 7493 activates a different set of outputs on the 7448. These outputs are tied to a resistor network that produces the control voltage for the voltage-controlled oscillator. The control voltage on pin 11 of the audio oscillator half of the 556 in combination with the setting of R12 and C2 will determine the frequency of the tones that the speaker will produce.

By experimenting with different combinations of resistor values for R3 to R9, you will produce a variety of different tone combinations. Varying R1 will change the rate at which the 7493 will cycle through its count. I used the following values. They produced a pleasing sequence of tones when R1 was adjusted to cycle quickly.

R3	15000	Ω
R4	1200	Ω
R5	22000	Ω
R6	1200	Ω
R7	100	Ω
R8	27000	Ω
R9	4700	Ω

The truth tables for the 7493 and the 7448 are shown in Table 5-2. The pinout for the 556 was shown in Fig. 4-22. The pinout for the 7493 was shown in Fig. 5-9.

COUNT	7493 OUTPUTS				7448 OUTPUTS						
	d	c	b	a	a	b	c	d	e	f	g
0	0	0	0	0	1	1	1	1	1	1	0
1	0	0	0	1	0	1	1	0	0	0	0
2	0	0	1	0	1	1	0	1	1	0	1
3	0	0	1	1	1	1	1	1	0	0	1
4	0	1	0	0	0	1	1	0	0	1	1
5	0	1	0	1	1	0	1	1	0	1	1
6	0	1	1	0	0	0	1	1	1	1	1
7	0	1	1	1	1	1	1	0	0	0	0
8	1	0	0	0	1	1	1	1	1	1	1
9	1	0	0	1	1	1	1	0	0	1	1
10	1	0	1	0	0	0	0	1	1	0	1
11	1	0	1	1	0	0	1	1	0	0	1
12	1	1	0	0	0	1	0	0	0	1	1
13	1	1	0	1	1	0	0	1	0	1	1
14	1	1	1	0	0	0	0	1	1	1	1
15	1	1	1	1	0	0	0	0	0	0	0

Table 5-2. Truth Table for the Melody Maker.

6
CMOS ICs

CMOS integrated circuits use field-effect transistors as the active elements instead of the bipolar transistors used in TTL ICs. The FETs are designed on the silicon in the same manner as discussed in Chapter 3. They are laid down on the silicon chip and interconnected as discussed in Chapter 4.

Both digital and analog functions have been implemented on CMOS ICs. This chapter introduces projects that use simple digital logic gates, more complex counters, a CMOS version of the 555, and some sophisticated special-purpose ICs. Some CMOS chips are direct replacements for the TTL digital ICs. These ICs start with the label 74C. These are exact pin-for-pin substitutions. Other CMOS digital ICs are given a 4000 series number.

Be sure to review the special handling precautions discussed in Chapter 1 before starting the projects in this chapter. Although CMOS ICs are rather durable, they can be destroyed by static electricity.

THEORY

An understanding of TTL ICs will go a long way in understanding CMOS ICs. If you understand the material in the last chapter, you should not have any trouble making the transition to CMOS. There are some important differences as well as advantages and disadvantages in using CMOS. These points are discussed in the following sections.

CMOS versus TTL

Probably the biggest advantages to using CMOS are the low power consumption and the wider choice of supply voltage. The projects will operate off of battery power much, much longer than the same circuit built with TTL ICs.

The wide range of supply voltages also allows the use of the common 9-volt radio battery as the supply. This is probably the simplest and least expensive way to power the projects.

The biggest disadvantages are the increased susceptibility to damage from electrostatic discharge and the higher cost. For this reason, the chips are packaged in conductive foam and should only be handled after you are grounded. Avoid touching the pins with your fingers.

Most CMOS ICs can be powered by any voltage from 3 volts to 15 volts. Recall that TTL ICs require a supply voltage between 4.75 volts and 5.25 volts. CMOS switching levels are a function of the supply voltage: the higher the supply voltage the greater the voltage separation between the 0 and the 1 state. This is an important advantage if the circuit under design will have to withstand high levels of noise. A wider separation between the 1 and 0 state decreases the susceptibility of incorrect switching due to stray voltage spikes.

Additionally, you must connect all CMOS inputs to either ground or the supply voltage. Unlike TTL, which will operate even with some of its inputs floating, CMOS ICs will operate erratically with unconnected inputs. Good design practice calls for all inputs to be connected regardless of the type of IC that you are using.

Mixing CMOS and TTL

CMOS and TTL ICs can be mixed in the same circuit under certain conditions. Most importantly, the power requirements for TTL must be respected. If CMOS and TTL are used together, the circuit must use a 5-volt supply. Some of the projects in this chapter will mix both CMOS and TTL devices.

Handling Precautions

CMOS ICs can be easily damaged by static discharges. Guarding against a static discharge is not terribly difficult. First, always store CMOS ICs with all the pins stuck into conductive foam or stick the pins into aluminum foil. Second, always turn off the power to the circuit before making or breaking a connection. This might be a tempting step to avoid when building a circuit on a breadboard, but the chip can be destroyed by any arc that would jump between the connections. Likewise, do not insert or remove a CMOS IC without removing the power first. Thirdly, do not handle the chips any more than necessary. Touching them only increases the chance of zapping them with any static buildup from your body. Low humidity and wool or synthetic fabrics are environments to avoid. Carpeting is especially prone to generating static.

Until a CMOS is soldered into place, it is very vulnerable. But once it is in the circuit, it is quite rugged and will withstand considerable abuse. These precautions are not difficult to follow and should protect the majority of the CMOS ICs that you will use.

No. 52: Pendulum Clock

In this project CMOS and TTL are mixed together. Note that the supply voltage is 5 volts. The circuit in Fig. 6-1 is straightforward. Two hex inverters from a 7404 are used as an oscillator. The pinout for the 7404 was shown in Fig. 5-5. They clock the 4017 CMOS decade counter at about 10 Hz. The counter divides the input frequency by 10 giving approximately a 1 Hz output. The other two inverters decode LED1 and LED6 and drive the speaker, which simulates a ticking clock. The 7402 NOR gates provide the decoding needed to give the visual impression of the pendulum swinging back and forth by lighting the appropriate LEDs. The pinout for the 7402 NOR gate is shown in Fig. 5-11.

The 7404 and 7402 ICs can be replaced with a 4049 CMOS inverting Hex buffer and a 4001 CMOS quad NOR gate, respectively. The pinouts for the 4049 and 4001 do not match the 7404 and the 7402. A 74C02 and a 74C04 will be exact pinout replacements. If all CMOS is used, the supply voltage can range from 3 to 15 volts. Note, however, that an appropriate dropping resistor will be needed for the LEDs if the supply is greater than 5 volts. They can only handle 5 volts, and operating them at 3 to 4 volts will extend their life significantly.

The outputs of the 4017 are internally decoded into a different output for each of 10 sequential inputs. Then the count begins over again. The NOR circuitry on the 4017 outputs does some additional decoding to give an up and down count. Table 6-1 provides a truth table of sorts to explain the LED pendulum action.

The circuit is not highly stable or highly accurate, but by playing with the values of C1 and C2, you should be able to use the pendulum in a decorative application. The LEDs should be mounted in a gentle arc pattern as shown in Fig. 6-1. It could be used with the egg timer project covered later in this chapter.

COUNT	4017 OUTPUT PIN	LED ON
1	3	1
2	2	2
3	4	3
4	7	4
5	10	5
6	1	6
7	5	5
8	6	4
9	9	3
10	11	2

Table 6-1. Truth Table for the Pendulum Clock.

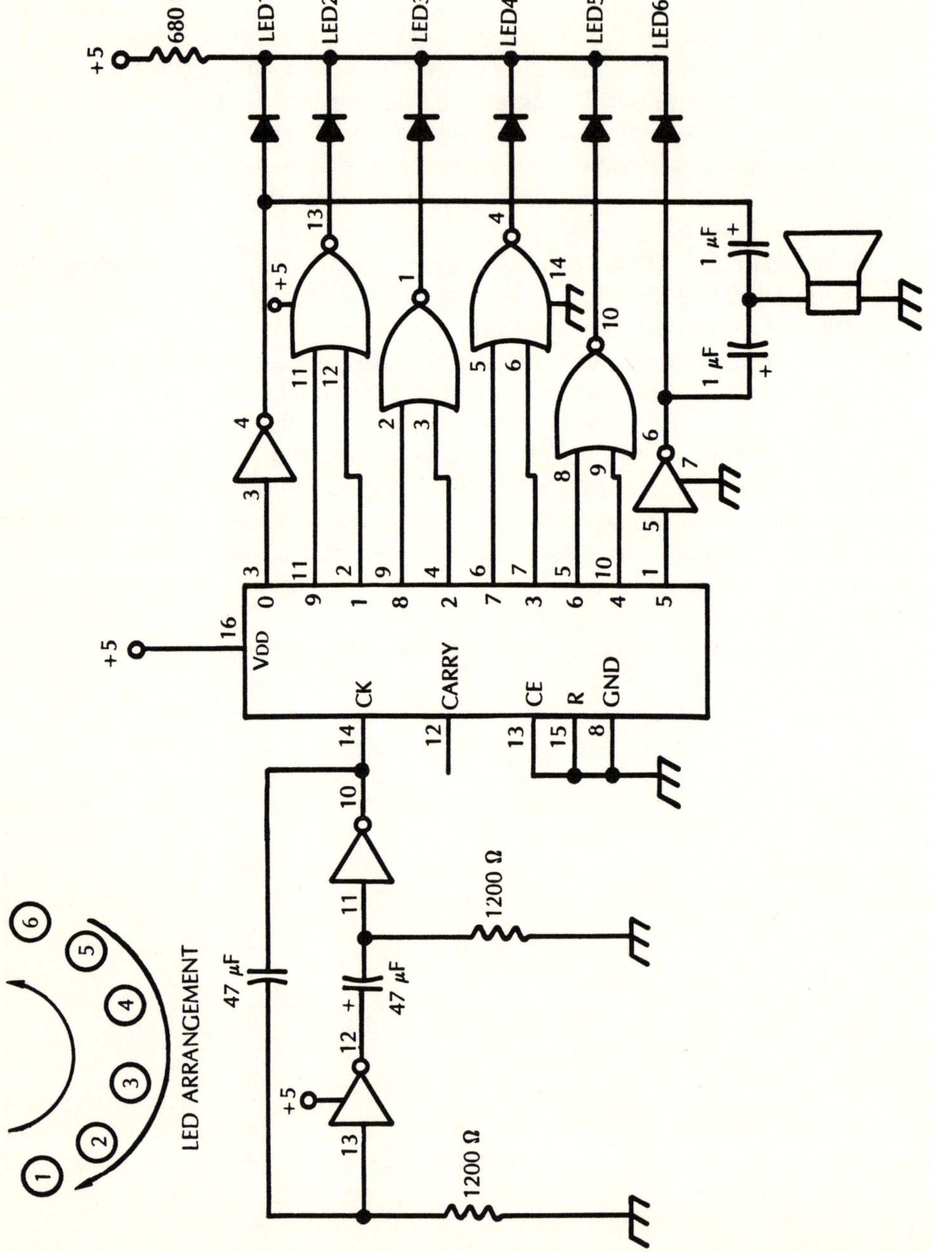

Fig. 6-1. Pendulum clock.

No. 53: CMOS Logic Probe

The logic probe in Fig. 6-2 uses one 4001 quad, dual-input NOR gate. It can indicate a 1, a 0, a pulse, and a floating input. By using the voltage supply for the circuit under test, it can indicate these four states accurately for supply voltages of 3 to 15 volts. The chart in Fig. 6-2 shows which LED indicates each state.

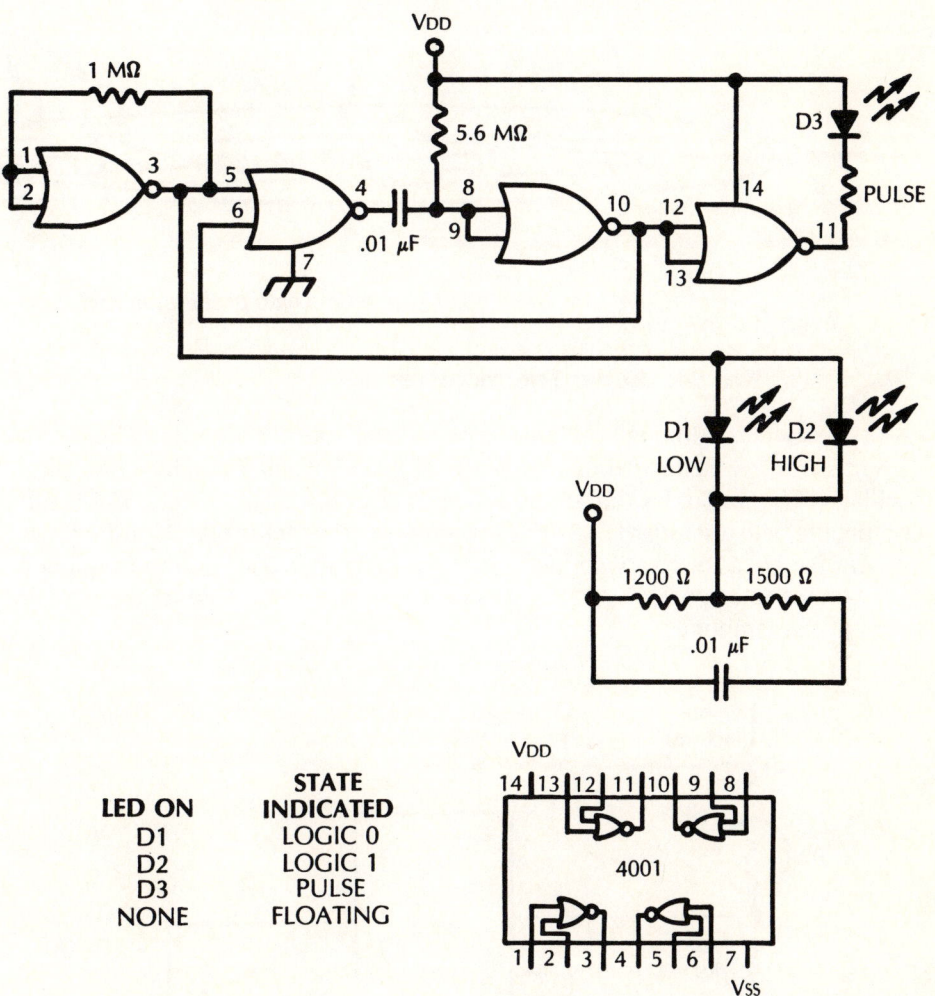

LED ON	STATE INDICATED
D1	LOGIC 0
D2	LOGIC 1
D3	PULSE
NONE	FLOATING

Fig. 6-2.

When connecting the probe to a point for testing, D3 may flash once quickly. Unless the input is a pulse, however, it will not stay lighted. If the frequency of the pulse is low (less than 10 Hz), D3 will pulse on and off visibly. When indicating a pulse, D1 and D2 may also light.

This probe will be very useful as a test device when troubleshooting new projects that you have built or for troubleshooting equipment that is already built. Construct it in a sturdy enclosure, possibly in the barrel of a wide ink pen. This will allow easy access to tight test points. Attach long leads for the V_{DD} and V_{SS} connections. A pair of alligator clips on each supply line will make for easy connections. The probe should be fairly long and pointed to allow an effective touch-down on a test point. If you are careful, you can solder component leads right to the IC pins. Refer to Fig. 6-3.

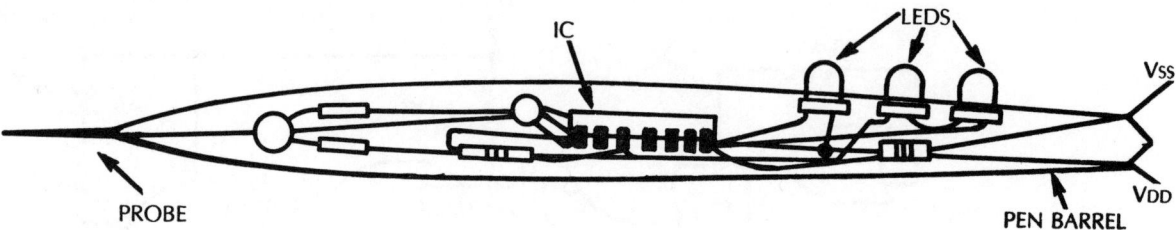

Fig. 6-3. CMOS Logic probe mounted inside a ball-point pen barrel.

No. 54: Audio Thermometer

Although of dubious practical value, the circuit in Fig. 6-4 illustrates the use of a CMOS 555 IC as a voltage-controlled oscillator. The pinout for the 555 is shown in Fig. 4-20. The CMOS 555 is pin compatible with a standard 555. The low current draw of the CMOS timer gives a 9-volt battery a long service life. Temperature is indicated by the frequency of the tone at the speaker. With some practice, you will be able to make a fair guess at the temperature.

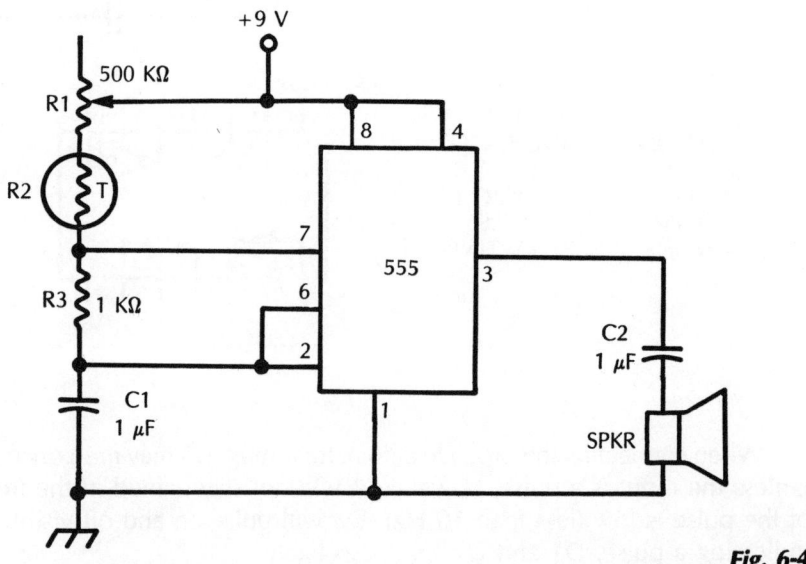

Fig. 6-4.

CMOS ICs 107

The thermistor (R2) changes its resistance value based on its temperature. Choose a thermistor that has a high resistance to temperature ratio. The 2D302 thermistor from NL Industries has a 500 Ω change between 32° F and 72° F (freezing and room temperature). This resistance differential is easily discernible for small changes in temperature.

No. 55: Random Number Generator

The 4017 counter can be used to generate random numbers for your project applications. Figure 6-5 shows a simple circuit to indicate a random number from 0 to 99. By cascading more 4017 stages, larger numbers can be generated. By using only one stage 0 to 9 can be generated. The pinout for the 4017 CMOS counter is shown in Fig. 6-1.

Nineteen LEDs are connected to the sequential outputs of the 4017. The first counter will indicate the number for the ones column. The second counter will generate the number for the tens column. Note that the first output of the second counter is not connected to an LED (pin 3). If this output was connected, it would not be possible to get a count between 0 and 9. Because one output on the counter is always high, the tens column would always have a LED lighted. By masking the first output, a zero in the tens column is possible.

The clock input should be a high frequency input (10,000 Hz or greater is fine). A 74C04 oscillator similar to the one in Fig. 6-1 could be used as a clock if smaller capacitors are substituted to increase the frequency. To approach being random, a high-frequency clock is needed. The 555 oscillator in Fig. 6-4 can also be used as a clock for the 4017. The variable resistor should be adjusted for the highest frequency. Substituting a smaller capacitor will also increase the frequency.

No. 56: Egg Timer

A very simple timer can be constructed from ICs. The one shown in Fig. 6-6 can be used to time anything from several seconds to over five minutes. The TTL inverter oscillator generates the starting clock frequency (about 3 Hz with the components shown). The two 4017 CMOS counters function as a divide by 100 circuit. By taking the control voltage off of pin 1 of the second 4017, the time interval is about three minutes. When pin 1 goes high, the 555 is allowed to sound the speaker. As you can see, if the 555 is triggered from a different counter output, a different time interval is obtained. Pin 11 of the second 4017 will give the longest count (about 5 minutes). The actual component values will have an effect on the timing in your circuit.

The 555 is connected in a simple astable configuration. When the control voltage on pin 5 is low, the oscillator is shut down. When pin 5 goes high the oscillator drives the speaker. Adjust the variable resistor (R3) to silence the speaker when pin 5 is low. R4 can be increased or decreased as desired for a more pleasing tone.

108 Beyond the Transistor: 133 Electronics Projects

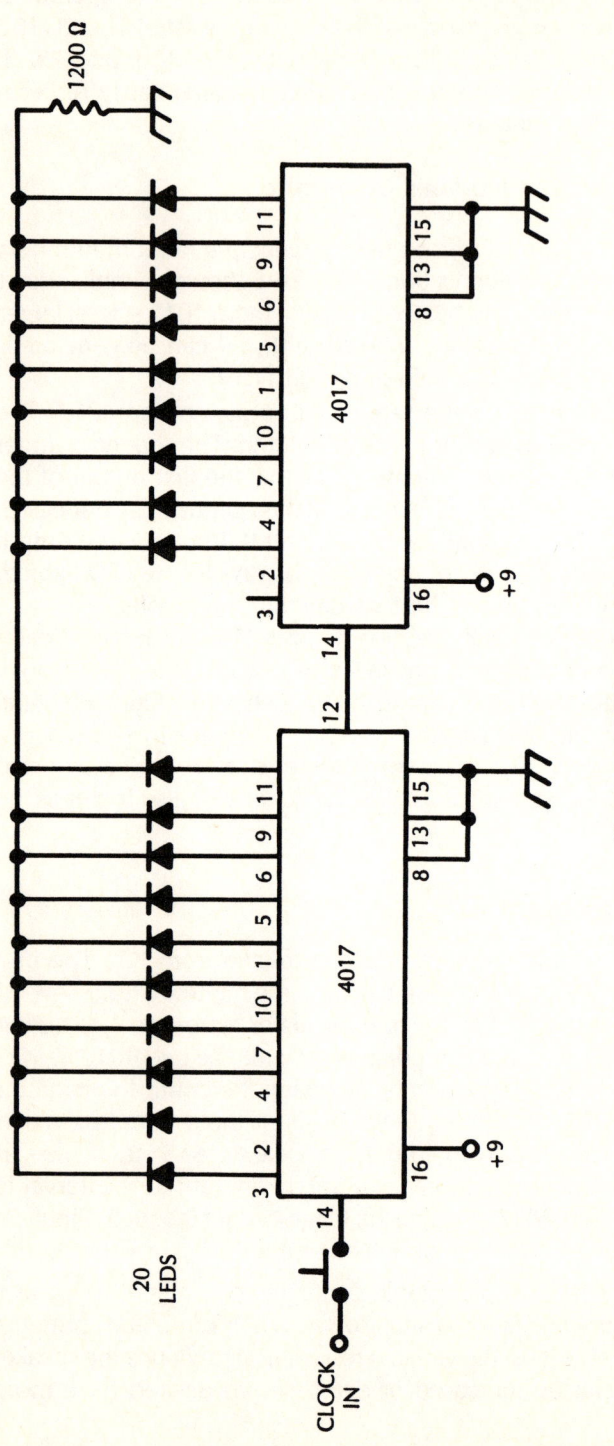

Fig. 6-5. *0 to 99 counter.*

CMOS ICs 109

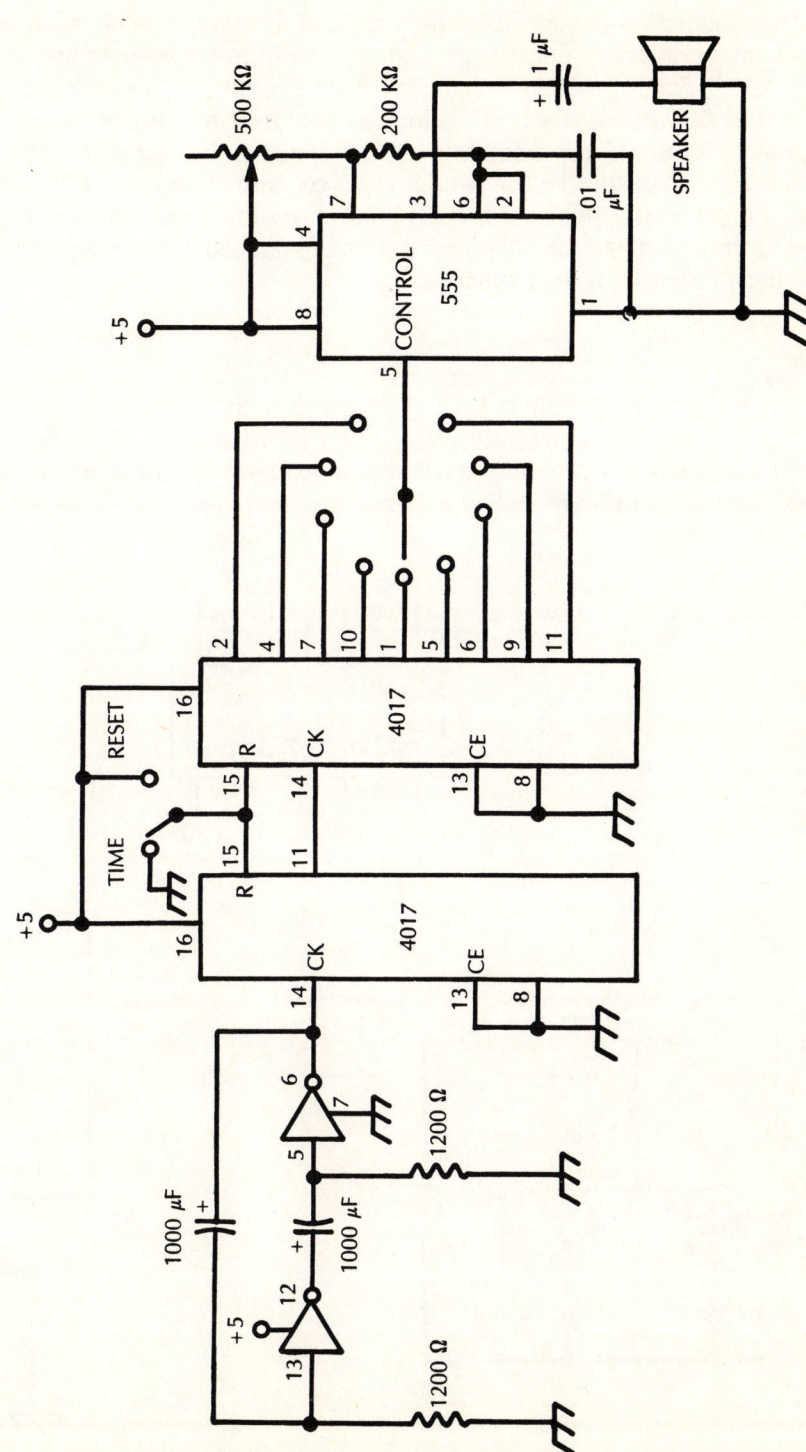

Fig. 6-6. Egg timer.

The multiple position switch is only needed if you want to have a multiple interval timer. If three minute eggs is all you will time, eliminate this switch. Wire the 555 directly to the pin that gives you the proper interval. You can also add an on/off switch and an LED to indicate that the timer is operating. If you like, the LED can be connected to the clock input of the first 4017. This way it will flash with the clock. The pendulum circuit shown in Fig. 6-1 could also be used in this circuit. Because an LED loads down the circuit, however, it will decrease the timing intervals. Although the timer is not highly accurate, for simple, noncritical timing, it is very suitable.

No. 57: Siren

The circuit in Fig. 6-7 can produce the two-tone European police siren sound that you have heard in the movies. The CMOS NAND gate oscillator produces the high/low pattern. A pinout for the 4011 NAND gate is also shown in Fig. 6-7.

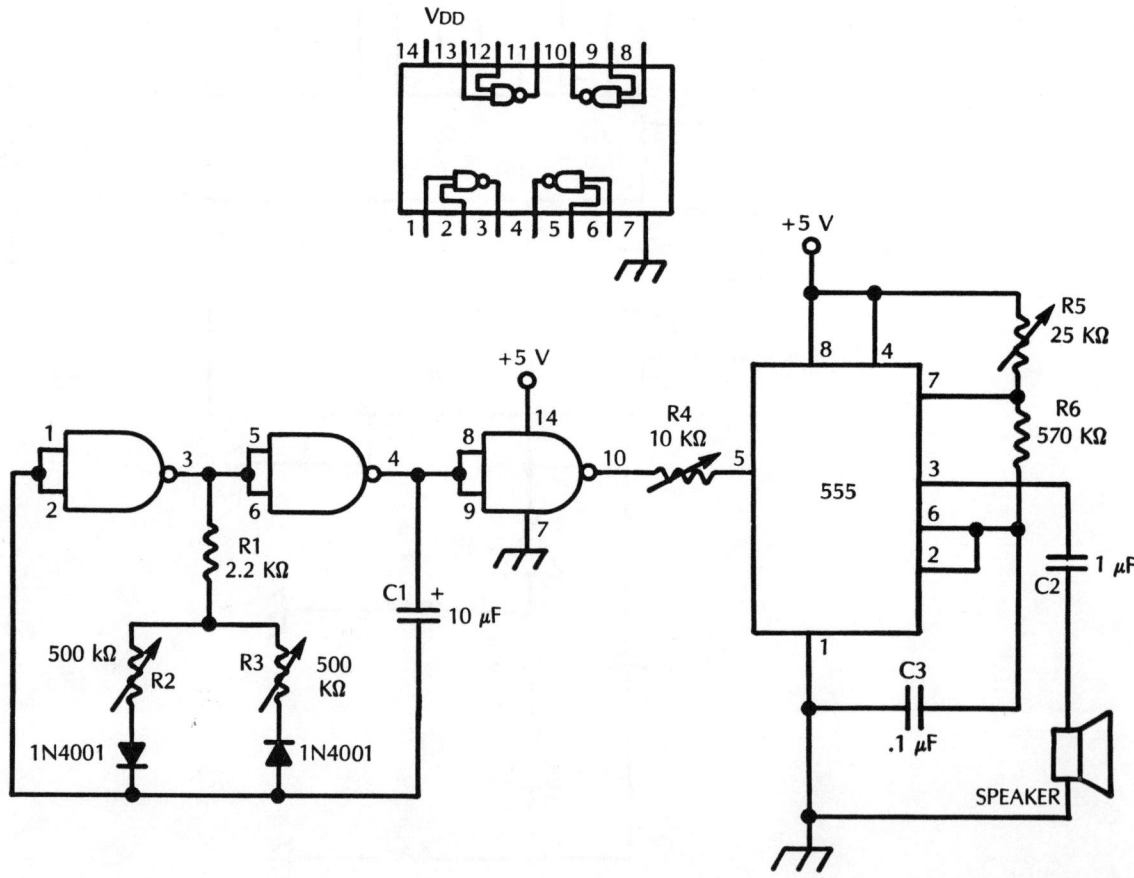

Fig. 6-7.

The length off each tone is controlled by R2 and R3. R5 determines the frequency of the tones. R4 is a tone control. Adjust these variable resistors until you get a sound that suits you. The actual component values that you use will have an effect on the sound that you get.

If you want a fancy alarm for the Egg Timer project, you can use this 555 circuit in the previous timer circuit. You will need to use the fourth NAND gate on the 4011 IC to turn the alarm on because the control voltage input on the 555 is used to produce the two tone sound. A CMOS oscillator will also need to be used unless you run the siren circuit at 5 volts.

No. 58: Metronome

The metronome circuit in Fig. 6-8 can function off of a single 9-volt battery for over a year. This circuit again shows the significance of the low power drain of CMOS ICs. It can also be made incredibly small for music practice on the run. The volume of the tick-tock sound is not very loud, but it should suffice if the metronome is located physically close to the user.

The metronome has separate rate and tone controls. The circuit in Fig. 6-6 could be adapted for use as a more sophisticated metronome. The oscillator frequency can be increased by replacing the 1000 μF capacitors with smaller ones. The different switch positions will provide set increments for the user. The 555 circuit could be replaced with the second half of Fig. 6-8. The switch would feed NOR gates C and D instead of the 555 circuit.

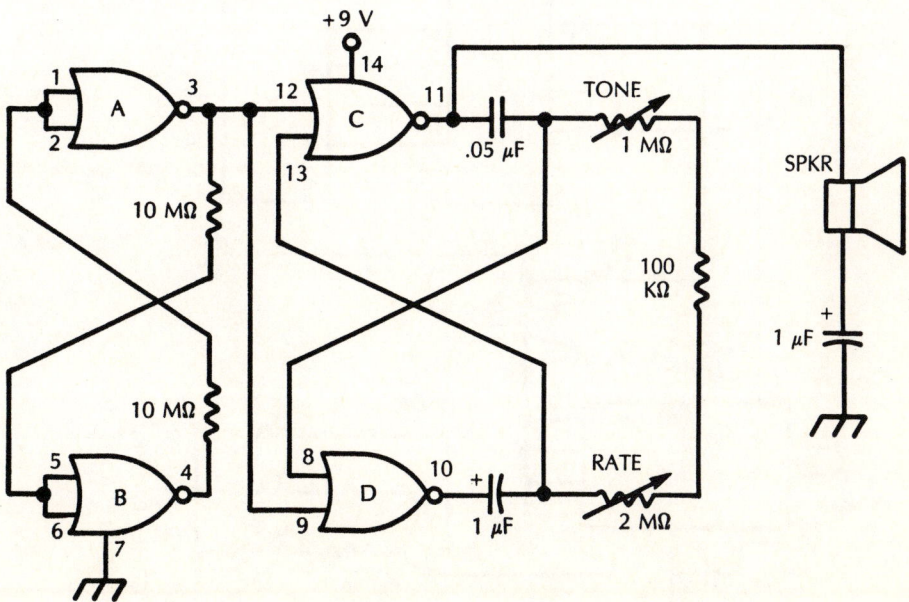

Fig. 6-8.

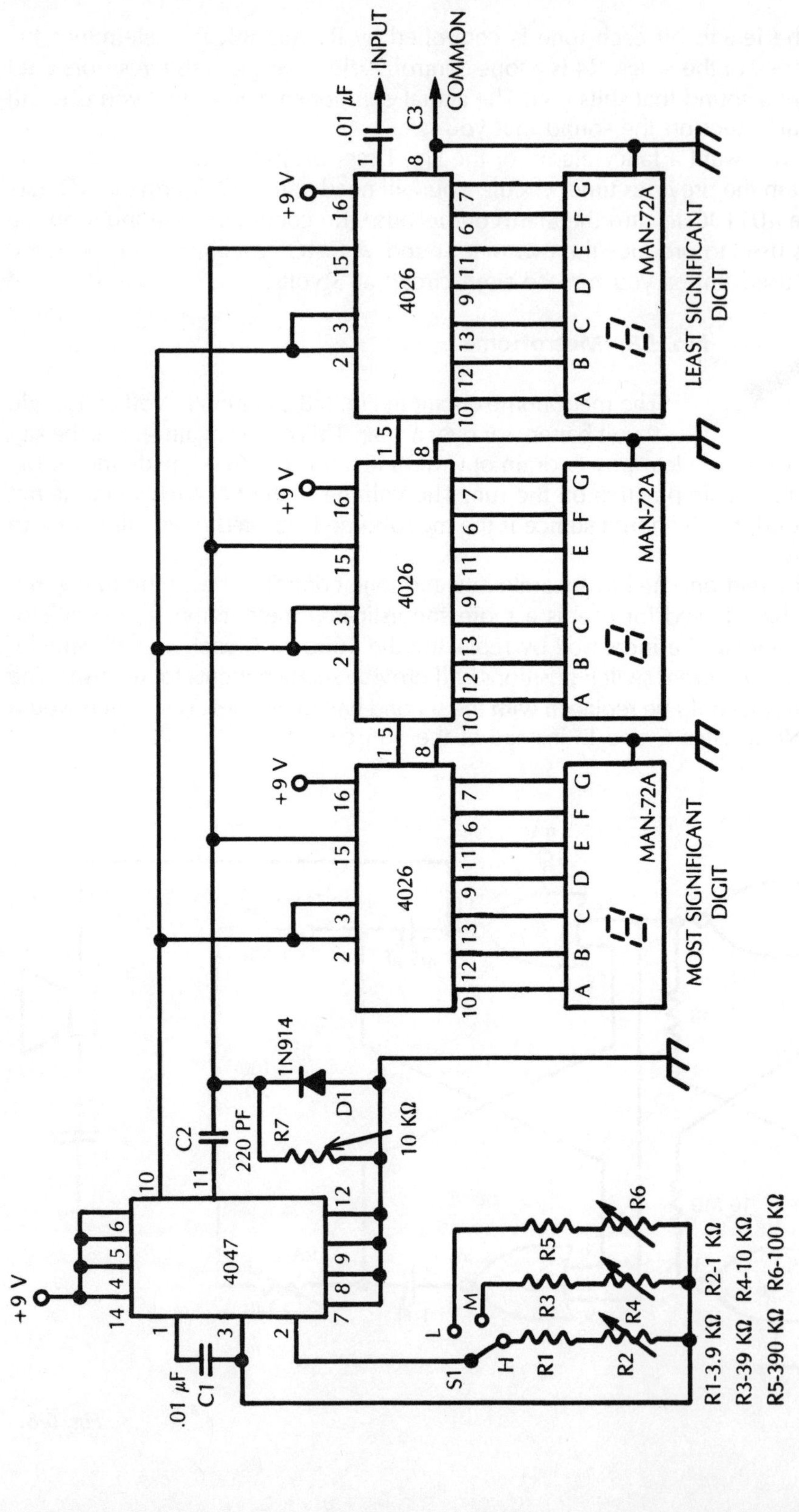

112 Beyond the Transistor: 133 Electronics Projects

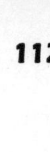

Fig. 6-9. Frequency counter.

No. 59: Frequency Counter

The simple frequency counter in Fig. 6-9 can detect frequencies up to about 4 MHz. Because it only has a three-digit display, high frequencies cannot be read very accurately. The simple design is a result of a high level of integration on the ICs that were chosen. The 4047 multivibrator generates the blanking and display pulses. The 4026 ICs are modulo-10, five-stage ring decade counters with a decoder driver for a seven-segment display also on the chip. The cascaded 4026 ICs simplify the design greatly by incorporating the display circuitry onto the chip. Figure 6-10 shows the pinouts for the 4026 and 4047.

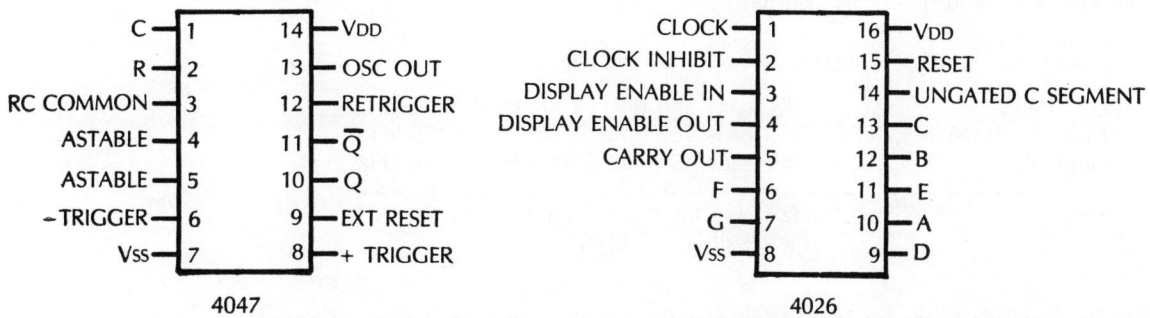

Fig. 6-10. Pinouts for the 4026 and 4047 ICs.

The 4047 is set up as a multivibrator running at a frequency set by C1 and the resistance of the variable resistor for the respective range. Note that the value of R1, R3, and R5 are each a decade larger than the previous one. The multivibrator thus will run at one frequency on range 1, ten times faster on range 2, and 100 times faster on range 3.

A signal on the input will increment the counters as long as pin 2 (clock inhibit) of the 4026 is low. Pin 3 is the display enable. A low on this pin blanks the display. Because pin 2 and 3 are tied together, the display is blanked while the counter is counting. When the clock inhibit goes high, the counters stop counting. A high on the display enable unblanks the display, and the current count is displayed. By connecting the reset pin of the 4026 (pin 15) to the NOT Q output, the counter is reset just before the next count cycle begins.

Calibration

Calibration of each range of the frequency counter is necessary before it can be used. There are several ways that this can be accomplished. The best solution is to obtain access to a frequency generator. By using a known frequency input that falls somewhere near the midrange of each range, adjust the variable resistor for the correct reading. Another method would be to use a second frequency counter that is already calibrated. Then any signal source can be used.

Your counter readout is simply adjusted to match the one that you borrowed.

A third method is to use a crystal controlled oscillator to adjust the frequency. Figure 2-6 is just such a circuit. One last calibration method is to use some readily available frequency standard. For instance, The i-f frequency for the AM band is 455 Hz. By tapping off this signal, you can get a frequency standard.

Reading the Counter

Using the frequency counter is a little tricky. Whenever a frequency is sampled, always check the frequency on all three ranges. By combining the information at all three ranges, you should be able to arrive at a good estimate of the frequency. For example, if you were to read the i-f frequency of an AM receiver, you would see the following:

```
                  MSD┐  ┌LSD
Low Range:                    This reads 550 times 100 Hz or 55 kHz.
Medium Range:     455 000     This reads 455 times 1000 Hz or 455 kHz.
High Range:       045 0000    This reads 45 times 10000 Hz or 450 kHz.
--------------------------------------------------------------
Result:           455,000 Hz  Using the digits from all three ranges,
                              the result is 455 kHz.
```

The reading from the low range overflowed the counter. The reading on the medium range gave a three digit count. The high range did not fill up the third digit so the measured frequency must be less.

The clock frequency of a personal computer is another standard that can be used. Consider the following reading from the clock of the TI 99/4A:

```
                  MSD┐  ┌LSD
Low Range:        998 00      This reads 998 times 100 Hz or 99.8 kHz.
Medium Range:     899 000     This reads 899 times 1000 Hz or 899 kHz.
High Range:       289 0000    This reads 289 times 10000 Hz or 2.89 MHz.
--------------------------------------------------------------
Result:           2,899,800   The actual clock frequency is 3 MHz.
```

The counter overflowed on both the low and medium ranges. The high range would look to have three good digits. When measuring high frequencies, it is possible to overflow the third range and get an incorrect reading. In this case, the actual frequency was already known, so the interpretation was safe. Note that the measured frequency was within about 1 percent of the actual.

No. 60: Music Box

The final project in this chapter uses a CMOS IC that has both digital and analog circuitry on the chip. The chip is very complex. It contains a dedicated microprocessor, Read-Only Memory, digital decoding circuitry, a digital-to-analog converter, and a system clock. All of this (and more) is contained in one 28-pin package. The pin-out of the AY-3-1350 Music Synthesizer chip is shown in Fig. 6-11.

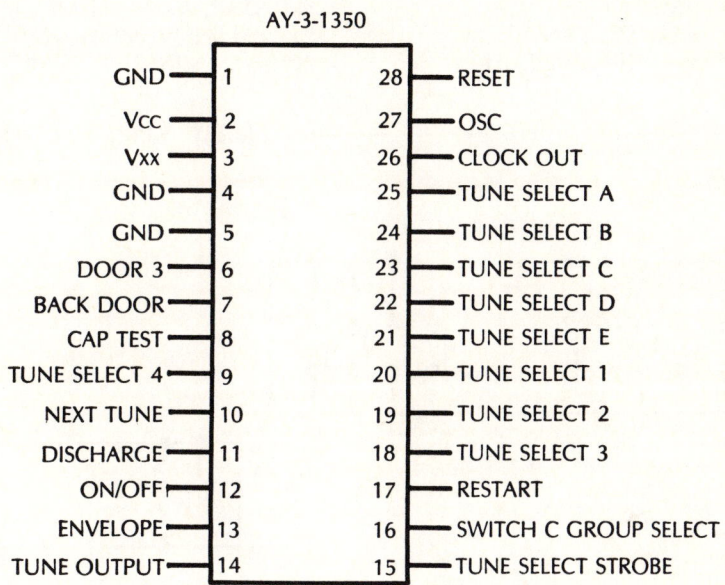

Fig. 6-11. Pinout for the AY-3-1350.

The AY-3-1350 holds 26 tunes in ROM. Each tune can be accessed by a different two-digit code. The names of the tunes and their codes are listed in Table 6-2. The tunes vary from old favorites to classical music. Even the recent *Stars Wars* theme has been etched into silicon.

Even with all the integration on the chip, this project still has a fair amount of external parts. Figure 6-12 shows the schematic for the music box circuit. The external circuitry breaks down into three general parts. The switches at the top control tune selection. The digital input to these pins is decoded into the address for each tune that is stored in ROM. The codes in Table 6-2 that have a zero as the second digit of the code do not require a switch to be pressed for the second digit. Just hold the proper letter and press the play switch. The last tune, Westminster Chimes, does not require any tune selection. Just press the play button with no other switches depressed.

Beyond the Transistor: 133 Electronics Projects

A0	TOREADOR	D2	BEETHOVEN'S FIFTH
B0	WILLIAM TELL OVERTURE	E2	AGUSTINE
C0	HALLELUJAH CHORUS	A3	O SOLE A MIO
D0	STAR SPANGLED BANNER	B3	SANTA LUCIA
E0	YANKEE DOODLE DANDY	C3	THE END
A1	JOHN BROWN'S BODY	D3	BLUE DANUBE WALTZ
B1	CLEMENTINE	E3	BRAHM'S LULLABY
C1	GOD SAVE THE QUEEN	A4	HELL'S BELLS
D1	COLONEL BOGEY MARCH	B4	JINGLE BELLS
E1	LE MARSEILLAISE	C4	LA VIE EN ROSE
A2	AMERICA THE BEAUTIFUL	D4	STAR WARS THEME
B2	DEUTSCHLAND LEID	E4	BEETHOVEN'S NINTH
C2	WEDDING MARCH	00	WESTMINSTER CHIMES

Table 6-2. Names and Codes.

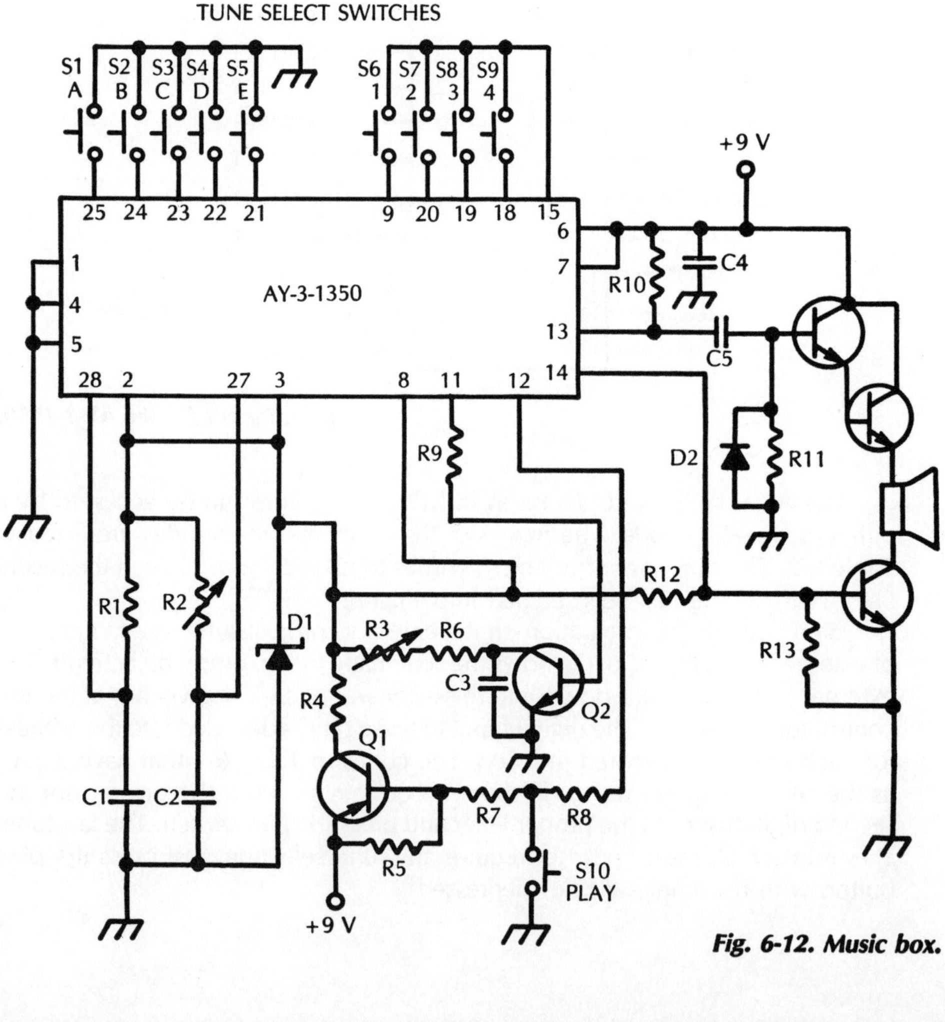

Fig. 6-12. Music box.

The circuitry to the right of the chip is the output circuitry. Because the chip does not have an amplifier integrated on the silicon, you must supply one to drive the 8 Ω speaker. The circuitry at the bottom of the chip controls power distribution and timing. The parts list for this circuit is given in Table 6-3. All of the parts are readily available at Radio Shack. The circuit will operate from a 9-V, transistor-radio battery. R2 is the pitch control. Adjust it to suit your own musical tastes. R6 is the speed control and controls the tempo of the music. For example, you can adjust this control to give a waltz a different tempo than a march.

This project demonstrates more than any other in this book the power of integration. To construct the music box circuit entirely from discrete components would be a rather formidable task. It also demonstrates the simplicity of ICs when they are treated as black boxes.

R1	100 kΩ
R2	25 kΩ POTENTIOMETER
R3	1 MΩ POTENTIOMETER
R4	33 Ω
R5	10 kΩ
R6	330 kΩ
R7	3.3 kΩ
R8	3.3 kΩ
R9	10 kΩ
R10	3.3 kΩ
R11	47 kΩ
R12	2.7 kΩ
R13	33 kΩ
C1	.1 µF
C2	47 pF
C3	.22 µF
C4	10 µF
C5	10 µF
D1	1N4735, 6.2 V ZENER
D2	1N4001
Q1	MPS 2907 pnp
Q2-Q5	MPS 2222 npn
S1-S10	spst, NORMALLY-OPEN MOMENTARY

Table 6-3. Parts List for the Music Box.

7
UJTs

The unijunction transistor (UJT) is a 3-terminal device which, unlike the bipolar transistor, has only one pn junction. Because the UJT exhibits negative input resistance, it is useful in a number of low-voltage devices—such as multivibrators, sawtooth generators, threshold detectors, pulse generators, and timers—which can exploit negative resistance and are simplified by use of the UJT instead of some other component.

The UJT is small and versatile and is manufactured in a number of types to meet various current and voltage demands. Although it functions most often by itself in such circuits as oscillators, the UJT is also found operating in conjunction with other devices such as silicon controlled rectifiers, triacs, and bipolar transistors.

THEORY

In the unijunction transistor (see Fig. 7-1A), an n-type silicon bar is provided with an ohmic base connection at each of its ends. These are designated base 1 and base 2. A pn junction is created in the bar near base 2 and is designated the emitter (E). This junction acts simultaneously as an emitter and as a collector when the UJT is in operation. When the device is biased in the manner shown in Fig. 7-1A, the portion of the bar between base 1 and the junction provides the emitter action, whereas the portion between the junction and the base 2 acts as a collector. This action is detailed in the following paragraphs.

When the switch is open, current I_{B2} from dc supply V_{BB} flows—from base 1 to base 2—through the silicon bar. In this state, the bar acts only as a resistance, and an approximately linear voltage gradient is set up; that is, the voltage drop between base 1 and any point along a straight line between the bases is proportional to the distance from base 1 to that point. The voltage opposite the pn junction is higher than ½V_{BB}, since the junction is more than halfway along the bar.

UJTs **119**

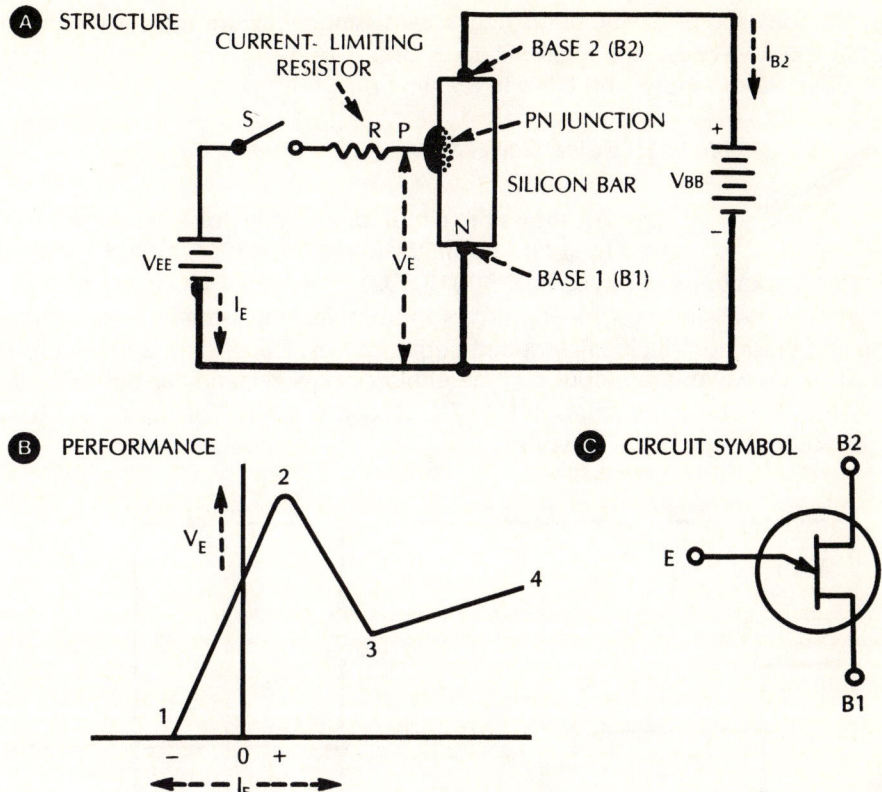

Fig. 7-1. Details of a unijunction transistor.

The dc supply V_{EE} is poled such that the p part of the emitter junction is positive with respect to base 1. When the switch is closed, forward current, therefore, flows through the junction, and holes are injected into the bar. Thus, the pn junction becomes an emitter. This hole current lowers the resistance of the bar between the emitter and base 1 and alters the voltage gradient.

If V_{EE} and V_{BB} are correctly proportioned, the p-region then will be biased positive over its half nearest base 1, and this half will act as an emitter; and the p-region will be biased negative over its half nearest base 2, and this half will act as a collector. In this way, the single junction functions simultaneously as emitter and collector.

When emitter current I_E is varied by adjusting voltage V_{EE} from a small negative value, through zero, to the maximum allowable positive value, a response characteristic similar to Fig. 7-1B is obtained: The resulting emitter-to-base-1 voltage (V_E) increases with I_E from zero at point 1 to the peak point at 2, then decreases to the valley point at 3, and finally increases (as from 3 to 4) as I_E further increases. The portion of the curve between 2 and 3 shows negative resistance, since here an increasing current produces a decreasing voltage drop. It is this negative-resistance characteristic that suits the UJT to its classic applications. On each side of the negative-resistance region, there is a conventional, positive-resistance region (1 to 2, and 3 to 4).

Figure 7-1C shows the UJT circuit symbol.

No. 61: Pulse Generator

Figure 7-2 shows the circuit of a simple pulse generator comprised by a UJT oscillator (type 2N2420, Q1) and a silicon bipolar output transistor (type HEP 50015, Q2). The output voltage of the UJT, taken across 47-ohm resistor R3, drives the bipolar transistor between saturation and cutoff, producing flat-topped output pulses. Depending upon the time off (t) of the wave, the output may be either narrow rectangular pulses or (as shown at the output terminals in Fig. 7-2) a square wave. The peak amplitude of the output signal is +15 volts.

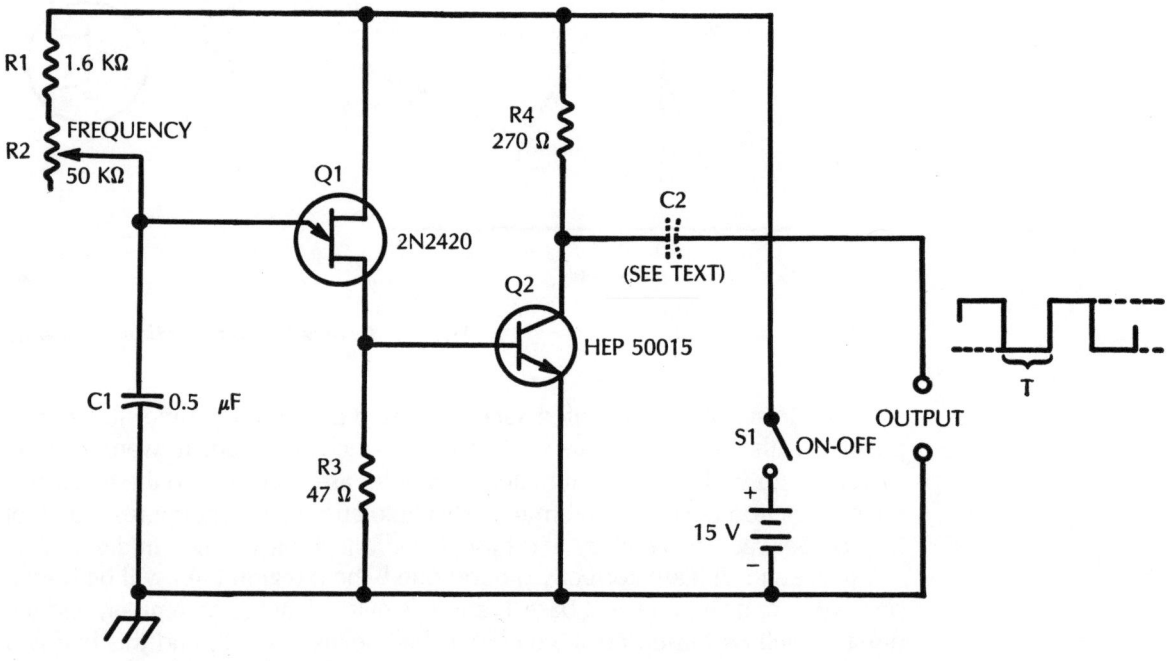

Fig. 7-2.

The frequency, or repetition rate, is governed by the setting of a 50,000-ohm rheostat and the capacitance of C1. At the maximum-resistance setting of this control, the full resistance R1 + R2 (i.e., 51.6 kilohms) is in the circuit with C1 = 0.5 μF. For this combination, the frequency (f) = 47.2 Hz, and the time off (t) = 21.2 ms.

At the minimum-resistance setting of R2, ideally only R1 (1.6 kΩ) is in the circuit; and for this condition, f = 1522 Hz, and t = 0.66 ms. For other frequency ranges, R1, R2, or C1—or all of them—may be changed:

$$t = 0.821 \, (R1 + R2) \, C1$$

where t is in seconds, R1 and R2 in ohms, and C1 in farads, and f = 1/t.

The circuit draws 20 mA from the 15-Vdc supply, but this value is subject to variation with individual UJTs and bipolars. The dc output coupling is shown in Fig. 7-2, but ac coupling may be provided by inserting a capacitor C2 in the high output lead, as shown by the dotted symbol. The capacitance of this unit should be between 0.1 and 1 μF, the best value being the one that causes least distortion of the output wave when the generator is operated into a particular desired load device.

No. 62: Pulse and Timing Generator

A simple sawtooth generator with sharp spikes is useful in a variety of applications involving timing, sychronizing, sweeping, and so on. UJTs generate such waveforms in simple and inexpensive circuits. Figure 7-3 shows such a circuit which, though not a precision instrument, will give a good account of itself in shops and low-budget laboratories.

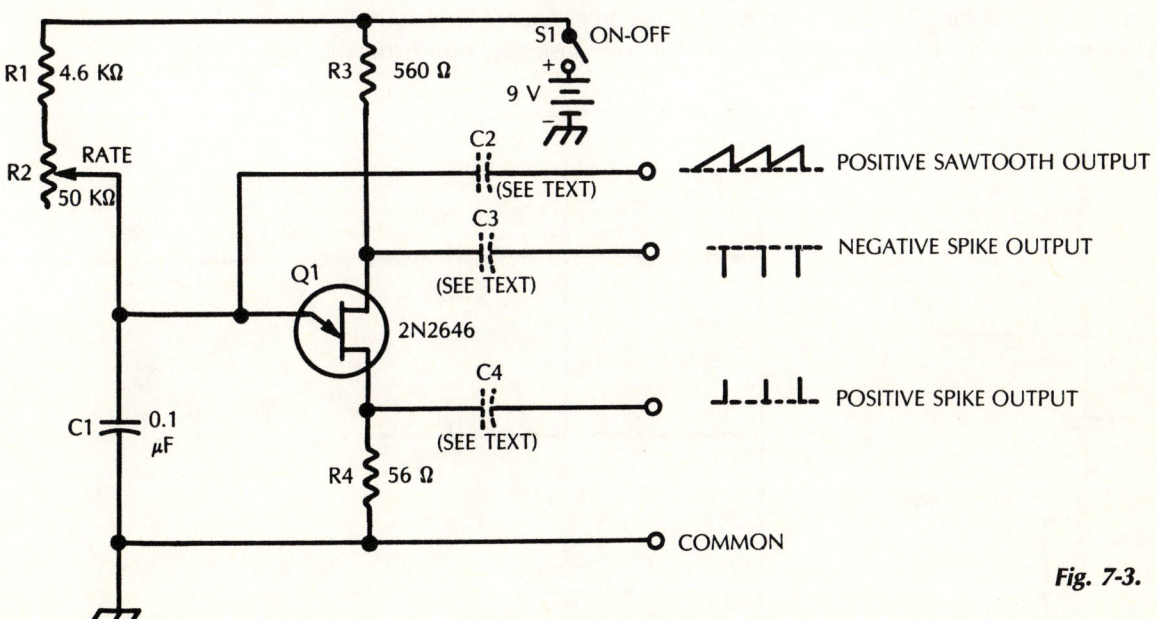

Fig. 7-3.

This circuit is essentially a relaxation oscillator, with outputs taken from the emitter and both bases. The 2N2646 UJT is connected in the classic oscillator circuit for these devices. The frequency, or repetition rate, is governed by the setting of the rate control rheostat, R2. When this control is set to its maximum-resistance point, the total resistance in series with timing capacitor C1 is the sum of the rheostat and the limiting resistance, R1 (that is, 54.6 kΩ), and the frequency is 219 Hz.

When R2 is set to its lower end, the total resistance ideally is that of resistor R1, or 5600 ohms, and the frequency is 2175 Hz. Other frequency limits and tuning ranges may be obtained by changing R1, R2, C1, or all three.

Positive-spike output is obtained from base 1 of the UJT, negative-spike output from base 2, and positive sawtooth output from the emitter. While dc output coupling is shown in Fig. 7-3, ac coupling may be obtained by inserting capacitors C2, C3, and C4 in the output leads, as shown by the dotted symbols. These capacitances will be between 0.1 and 10 μF, the value selected being determined by the maximum capacitance which can be tolerated with a given load device without degrading the output waveform.

The circuit draws approximately 1.4 mA from the 9-volt dc supply. All resistors are ½-watt.

No. 63: Free-Running Multivibrator

The UJT circuit shown in Fig. 7-4 is similar to the relaxation oscillator circuits described in the two preceding sections, except that its constants have been chosen to give quasi-square-wave output resembling that of a conventional astable multivibrator employing transistors or tubes.

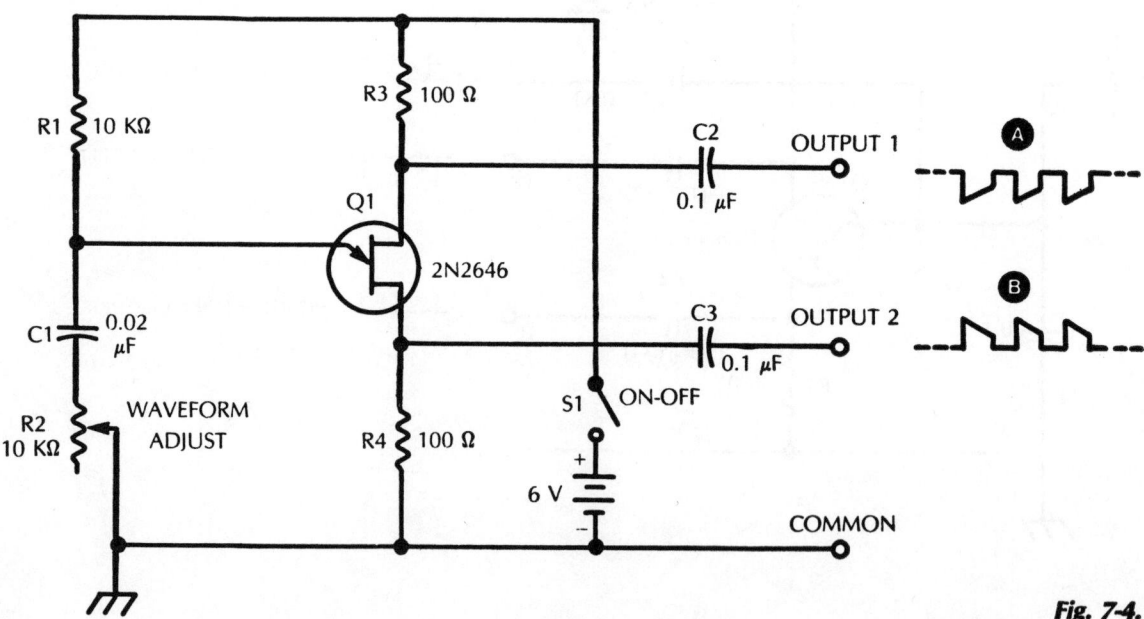

Fig. 7-4.

The type 2N2646 unijunction transistor is a good performer in this arrangement.

Two output signals are provided: a negative-going pulse at base 2 of the UJT (see Fig. 7-4A), and a positive-going pulse at base 1 (see Fig. 7-4B). The open-circuit peak amplitude of each of these signals is approximately 0.56-volt, but this may vary somewhat with individual UJTs. The 10,000-ohm rheostat, R2, must be adjusted for desired tilt or flatness of the top of the output wave. This control also affects the frequency, or repetition rate. With the values given here for R1, R2, and C1, the frequency is approximately 5 kHz for a flat-topped peak. For other frequencies, change R1 or C1:

$$f = 1/0.821\ RC$$

where f is in Hz, R in ohms, and C in farads.

The circuit draws approximately 2 mA from the 6-Vdc supply. All fixed resistors are ½-watt.

No. 64: One-Shot Multivibrator

Figure 7-5 shows the circuit of a one-shot multivibrator (this device is also called a single-shot multivibrator or monostable multivibrator). A type 2N2420 unijunction transistor and type 2N2712 silicon bipolar transistor are combined to give a single, constant-amplitude output pulse each time the circuit is triggered by an input pulse. This circuit is adapted from a design by W. Sowa and J. Toole.

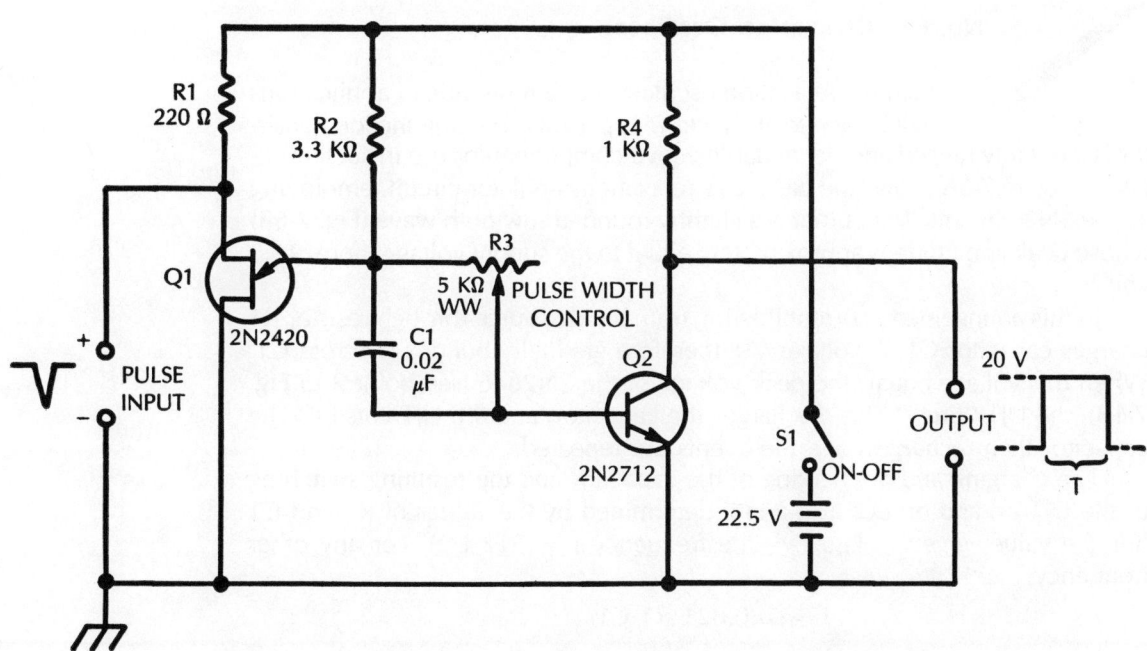

Fig. 7-5.

In this arrangement, the voltage divider formed by R2, R3, and the base-to-emitter resistance of transistor Q2 charges capacitor C1 making its Q2 end negative and its Q1 end positive. This divider also applies to the emitter of Q1 a positive voltage that is a little lower than the peak voltage of the 2N2420 (see point 2 in Fig. 7-1B).

Initially, Q2 is in the conducting state; therefore, the resulting voltage drop across resistor R4 reduces the voltage at the output terminals substantially to zero. When a 20-volt negative pulse is applied to the pulse input terminals, Q1 "fires," pulling the emitter side of C1 down to ground and biasing the Q2 base negative. This action cuts off Q1, and the Q1 collector voltage rises rapidly to +20 volts (see the pulse at the output terminals in Fig. 7-5).

The voltage remains at this level for the time t taken for capacitor C1 to discharge through resistor R3. The output then falls back to zero, and the circuit rests until another pulse arrives. Time t, and accordingly the width (duration) of the output pulse, depend upon the setting of the pulse width control, R3; with the values of R3 and C1 given in Fig. 7-5, the range is 2 μs to 0.1 ms, assuming that R3 covers the resistance range 100 to 5000 ohms. Other time ranges may be obtained by changing C1, R3, or both:

$$t = R3C1$$

where t is in seconds, R3 in ohms, and C1 in farads.

The circuit draws approximately 11 mA from the 22.5-volt dc supply. But this may be expected to vary somewhat with individual UJTs and bipolars. All fixed resistors are ½-watt.

No. 65: Relaxation Oscillator

A simple relaxation oscillator has a multitude of applications well known to all electronics persons. The unijunction transistor is a notably rugged and dependable active component for use in such oscillators. Figure 7-6A shows the basic UJT relaxation-oscillator circuit, employing a type 2N2646 unit. The output is a slightly rounded sawtooth wave (Fig. 7-6B) whose peak amplitude is approximately equal to the supply voltage (here, 22.5 volts).

In this arrangement, current flowing from the dc source through resistor R1 charges capacitor C1. A voltage V_{EE} therefore gradually builds up across C1. When this voltage equals the peak voltage of the 2N2646 (see point 2 in Fig. 7-1B), the UJT "fires." This discharges the capacitor, and the UJT cuts off. The capacitor then recharges, and the events are repeated.

The charging and discharging of the capacitor and the resulting switching of the UJT on and off occur at a rate determined by the values of R1 and C1 (for the values given in Fig. 7-6, the frequency f = 312 Hz). For any other frequency,

$$f = 1/(0.821\ R1\ C1)$$

where f is in Hz, R1 in ohms, and C1 in farads. A rheostat of suitable resistance may be substituted for the fixed resistor, R1, if continuously variable frequency control is desired.

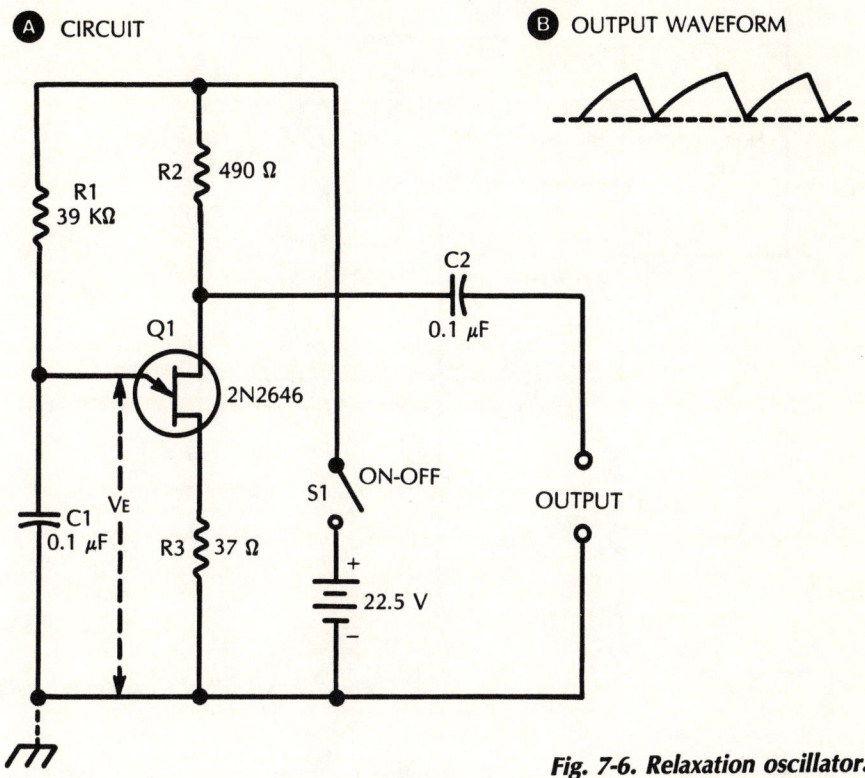

Fig. 7-6. Relaxation oscillator.

All resistors are ½-watt. Capacitors C1 and C2 may be any convenient low-voltage units; however, for good stability, C1 must be of good quality. The circuit draws approximately 6 mA from the 22.5-Vdc supply.

No. 66: Standard-Frequency Oscillator

Figure 7-7 shows the circuit of a 100-kHz crystal oscillator which is usable in the conventional manner as a secondary frequency standard or spot-frequency generator. This circuit delivers a distorted output wave which is very desirable in a frequency standard in order to insure strong harmonics high in the rf spectrum. The combined action of the HEP 310 unijunction transistor and the 1N914 diode harmonic generator produces this distorted wave.

In this arrangement, a small 100 pF variable capacitor, C1, allows the frequency of the 100 kHz crystal to be varied slightly, to bring a high harmonic, such as 5 MHz, to zero beat with a WWV/WWVH standard-frequency signal.

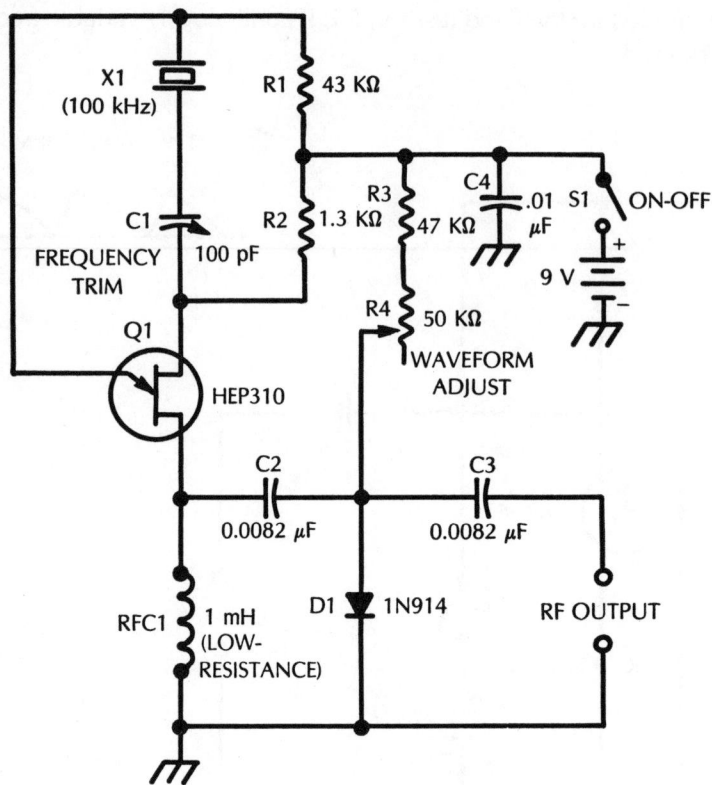

Fig. 7-7. Standard frequency oscillator.

The output signal is developed across the 1 mH rf choke (RFC1) which must have low dc resistance (J. W. Miller type 73F103AF has a resistance of only 7.5 ohms). This signal is applied to the 1N914 diode (D1) which is dc-biased through R3 and R4 to the most nonlinear part of its forward conduction characteristic, to further distort the UJT output. When the oscillator is being used, the waveform adjust rheostat, R3, is set for the strongest signal at the harmonic of 100 kHz being used. Resistor R3 serves only as a current limiter to prevent direct connection of the 9-volt supply across the diode.

The oscillator draws 2.5 mA from the 9-Vdc supply, however, this may vary somewhat with individual UJTs. Capacitor C1 must be a midget air-type; all other capacitors are mica or silvered mica. All fixed resistors are 1-watt.

No. 67: CW Monitor

The cw monitor circuit shown in Fig. 7-8 is powered directly by rf pickup from the transmitter being monitored, and delivers an adjustable-tone audio signal to high-impedance headphones. The volume of this tone depends upon the strength of the rf, but will be found adequate even with low-powered transmitters.

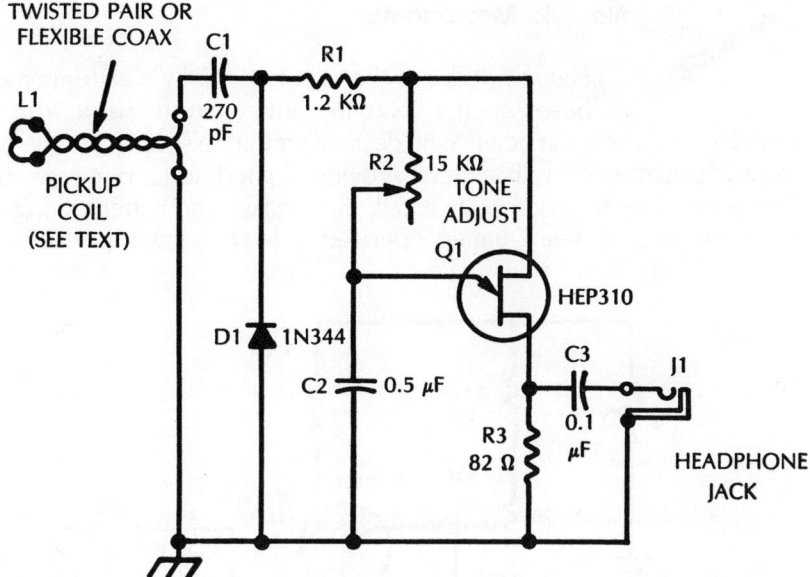

Fig. 7-8. Cw monitor.

The signal is sampled by the rf pickup coil, L1, which consists of 2 or 3 turns of insulated hookup wire mounted rigidly near the output tank coil of the transmitter. The rf energy is rectified by a shunt-diode circuit, consisting of blocking capacitor C1, diode D1, and filter resistor R1, and the resulting dc is used to power the HEP 310 unijunction transistor in a relaxation oscillator circuit. The output of this oscillator is applied to high-resistance headphones through coupling capacitor C3 and output jack J1.

The tone of the signal heard in the headphones may be adjusted over a good range by means of rheostat R2; the tone frequency is approximately 162 Hz when R2 is set to 15,000 ohms, and is approximately 2436 Hz when R2 is set to 1000 ohms. The volume may be controlled by moving L1 nearer to or farther from the transmitter tank; usually, a location will be found that gives satisfactory volume for all general use.

The device may be assembled in a small, grounded metal box. In general, it may be located at any reasonable distance from the transmitter, if good-grade twisted pair or flexible coaxial cable is used and if L1 is coupled to the low end of the tank coil. All fixed resistors are ½-watt. Capacitor C1 should be rated to withstand safely the maximum dc voltage that might accidentally be encountered in the transmitter; C2 and C3, however, may be any convenient low-voltage units.

This type of rf-powered monitor has some advantage over the usual sine-wave type in that the distorted signal delivered by the UJT circuit is often more pleasing to the ear over long periods than is the flute-like sinusoidal tone. The distorted tone is certainly less lulling.

No. 68: Metronome

Figure 7-9 shows the circuit of a fully electronic metronome based upon a 2N2646 unijunction transistor. This device is useful to musicians and others who desire a uniformly spaced audible beat. Driving a 2½-inch speaker, this circuit provides a good, loud, pop-type signal. The metronome can be made quite small, the speaker and battery being its largest components, and, being battery-operated, it is completely portable.

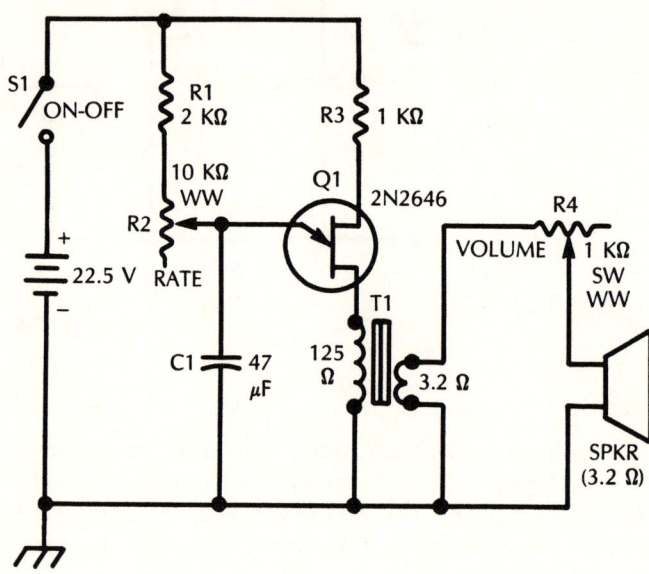

Fig. 7-9.

The circuit is a variable-frequency relaxation oscillator which is transformer coupled to the 3.2-ohm speaker. The beat rate is adjustable from approximately 1 per second (60 per minute) to approximately 10 per second (600 per minute) by means of a 10,000-ohm wirewound rheostat, R2. Volume is adjustable by means of a 1000-ohm, 5-watt, wirewound rheostat, R4. Output transformer T1 is a miniature 125:3.2-ohm unit (Argonne AR-174 with primary center tap unused, or equivalent). The circuit draws 4 mA at the slowest beat rate of the metronome and 7 mA at the fastest beat rate, but this can vary with individual UJTs. A 22.5-volt battery will give good service at this low current drain.

Electrolytic capacitor C1 is a 50-volt unit. Resistors R1 and R3 are ½-watt, and rheostats R2 and R4 are wirewound (R4 is 5-watts).

No. 69: Tone-Identified Signal System

The circuit in Fig. 7-10 enables a separate tone signal to be obtained from each of several stations. These stations may consist of different doors in a building, different desks in an office, different rooms in a house, or any other places at which pushbuttons may be installed. The place that is signaling can be recognized from its distinctive tone, provided there are not too many stations and that the tones are far enough apart (for example, 400 Hz and 1000 Hz) that they are easily distinguishable by ear.

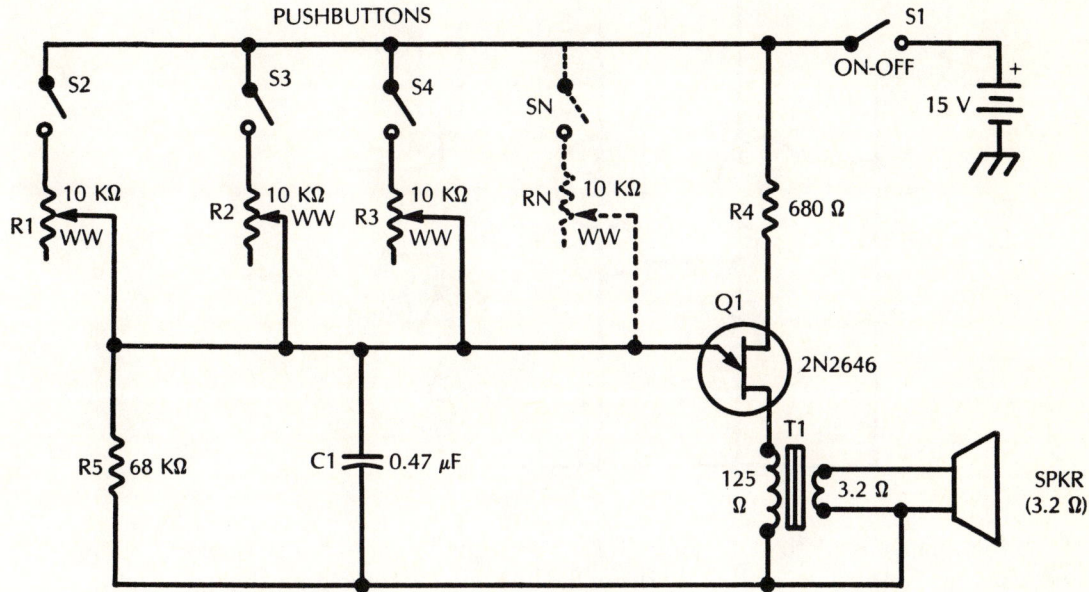

Fig. 7-10.

The circuit is a simple relaxation oscillator employing a type 2N2646 unijunction transistor to produce the signal and drive a loudspeaker. The tone frequency is set by capacitor C1 and one of the 10,000-ohm wirewound rheostats (R1 to R_n). When the rheostat is set to 10,000 ohms, the frequency is approximately 259 Hz; when the rheostat is set to 1000 ohms, the frequency is approximately 2591 Hz.

The oscillator is coupled to the speaker through output transformer T1, a miniature 125:3.2-ohm unit (Argonne AR-174 with primary center tap unused, or equivalent). The circuit draws approximately 9 mA from the 15-Vdc source, but this will vary with individual UJTs.

In this circuit, all fixed resistors are ½-watt. The capacitor may be any convenient low-voltage unit. The wires to the pushbuttons will add some capacitance to C1, but in most instances this will be negligible.

No. 70: Trigger For SCR

Figure 7-11 shows how a unijunction transistor may be used to trigger a silicon controlled rectifier. (See Chapter 13 for a discussion of silicon controlled rectifiers.) Here, a type HEP 310 UJT triggers a type HEP R1243 SCR.

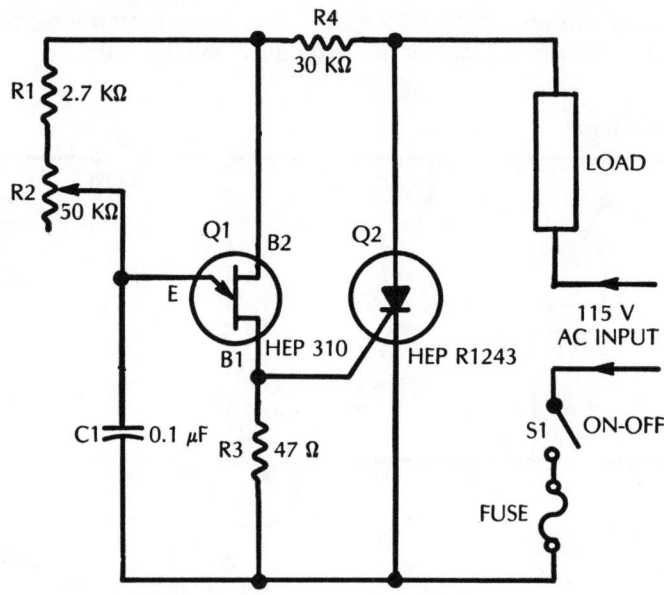

Fig. 7-11.

In this arrangement, the UJT and SCR both are supplied from the ac power line. Current flow through resistor R4 drops the line voltage sufficiently that the UJT operates within voltage specifications. The timing circuit (R1-R2-C1) likewise is operated from the line. The load may be a motor or any other device which is to be operated on the dc output of the SCR.

On the positive half-cycle of line voltage, Q1 "fires" at any instant determined by the time constant of the R1-R2-C1 section and the characteristics of the UJT and switches Q2 into conduction. At the end of this half-cycle, Q2 switches off because of the zero line voltage and remains off during the negative half-cycle. At the same time, capacitor C1 discharges through the internal emitter-to-base-1 path of the UJT. The circuit is then ready to repeat the events.

Adjustment of rheostat R2 allows selection of the instant (point in the positive half-cycle) at which the SCR turns on (the SCR then conducts until the end of the positive half-cycle and all during the negative half-cycle). This phase-controlled action allows easy adjustment of the output current flowing through the load.

In this circuit, all fixed resistors are 1 watt. Capacitor C1 must withstand safely the peak value of the line voltage, and for maximum safety and dependability should be a 600-volt unit.

8
VCDs

Fig. 8-1.

The variable-capacitance diode (VCD) is a junction diode that has been specially processed to make useful the inherent voltage-variable capacitance of reverse-biased pn junctions. The VCD therefore can act as a tiny dc voltage-tuned capacitor in various electronic circuits, such as LC tuners and automatic frequency control systems.

All reverse-biased semiconductor diodes exhibit voltage-variable capacitance, but this capacitance is very small in conventional diodes. In the VCD, capacitance is provided in useful amounts. The VCD is also called a varactor or a voltage-variable-capacitance diode (VVCD). It is also known by several trade names, such as Epicap, Semicap, and Varicap.

Because the VCD junction is reverse biased, it draws very little current (in some instances less than 1 nanoampere) from the dc control-voltage source, and, therefore, is to all practical purposes a voltage-controlled—i.e., "zero voltage"—device. Increasing the control voltage to its nominal value (as given by the VCD manufacturer) will decrease the capacitance to one-half its initial (zero-voltage) value. Up to a 10:1 capacitance change is readily obtainable at voltages higher than nominal, however. Depending upon make and model, the Q of a variable-capacitance diode is several hundred at frequencies of several hundred megahertz. Figure 8-1 shows the circuit symbol of the VCD.

THEORY

The variable-capacitance diode contains an abrupt pn junction. In every junction, there is a depletion layer about the junction, that is, a region in which no current carriers (or, at least, only a trace of them) are present. This layer thus is equivalent to a thin dielectric. The n- and p-regions, which are good conductors, face each other on opposite sides of the depletion layer and act as the two plates of a capacitor with the depletion layer acting as the dielectric of this capacitor.

When no external voltage is applied to the diode, the depletion layer is thin and the junction capacitance is highest (see Fig. 8-2A). But when a reverse dc voltage is applied to the diode (that is, anode negative and cathode positive), the depletion layer widens by an amount proportional to the voltage, and the capacitance decreases (see Fig. 8-2B). It is in this way that a variable dc voltage makes the diode a variable capacitor. When the junction is reverse biased, any direct current flowing through the diode from the control-voltage source is very small (usually in the nanoamperes), so virtually no power is required to tune this semiconductor capacitor.

A ZERO APPLIED VOLTAGE

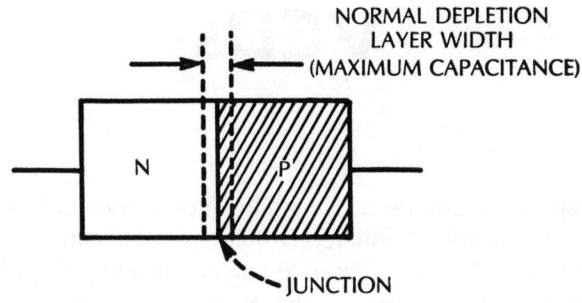

B HIGH APPLIED VOLTAGE

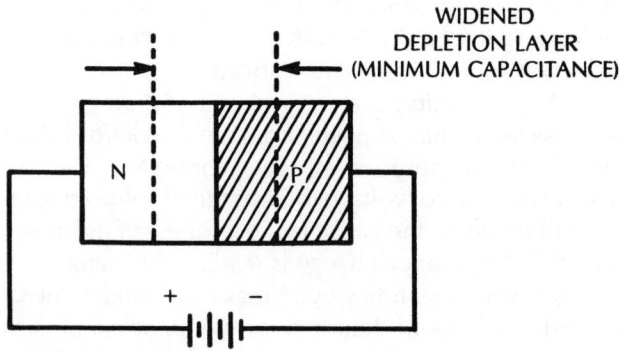

C VCD PERFORMANCE

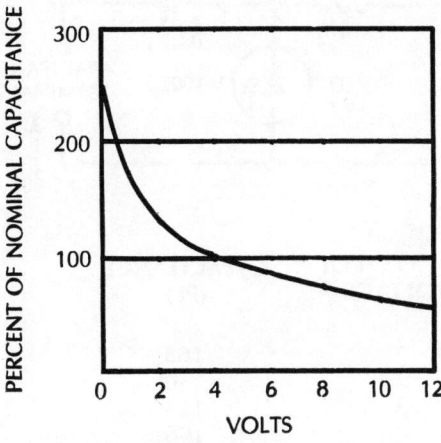

Fig. 8-2. VCD action.

Figure 8-2C shows typical performance of the VCD. From this plot, notice that an increase in voltage from zero to 1½ times the nominal value decreases the capacitance from approximately 260% of nominal capacitance value to approximately 80% of nominal capacitance. This means that a VCD rated at 100 pF at 4 volts would show a capacitance change from 260 pF at zero volts to 80 pF at 6 volts. Common nominal capacitance values for VCDs are given in steps from 6.8 pF to 100 pF, although values of several hundred picofarads are sometimes manufactured. A common nominal voltage is 4 Vdc, and maximum permissible voltages are given in steps from 15 Vdc to 150 Vdc.

No. 71: Voltage-Variable Capacitor

When an adjustable dc voltage is available from any source, it can be used to vary the capacitance of a variable-capacitance diode, and a dc-tuned miniature variable capacitor results. Such a component has a great many applications in communications and general electronics. Any VCD may be employed in this manner, the unit chosen being selected for desired voltage and capacitance ratings.

Figure 8-3A shows the circuit of a voltage-variable capacitor offering a 4:1 capacitance range. The capacitance seen at the capacitance terminals varies from approximately 260 pF to 65 pF as the dc control voltage varies from zero to +12 volts. Figure 8-3B details this performance.

In this circuit, the 1-megohm, ½-watt resistor (R1) isolates ac energy applied to the capacitance terminals in applications in which the voltage-variable capacitor is used, keeping it out of the dc voltage source. This resistor replaces the usual choke coil in this function; and since the VCD draws virtually no current in its reverse-biased condition, there is practically no voltage drop across R1.

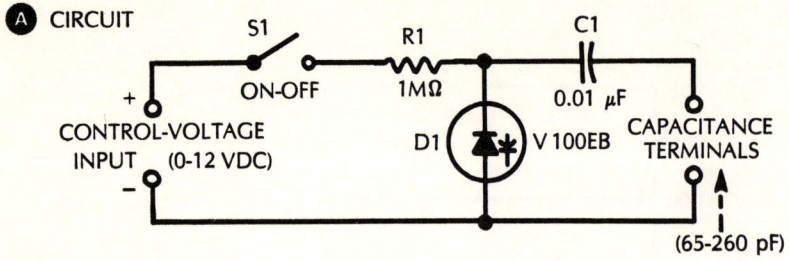

A CIRCUIT

B PERFORMANCE

DC CONTROL VOLTAGE	CAPACITANCE (PF)
0	260
1	165
2	130
3	115
4	100
5	92
6	85
7	80
8	79
9	72
10	69
11	68
12	65

Fig. 8-3. Voltage-variable capacitor.

The 0.01 μF capacitor (C1) protects the V100EB variable-capacitance diode (D1) from any dc component in the external circuit in which this arrangement is used. Since the capacitance of C1 is large, with respect to that of diode D1, the capacitance seen at the capacitance terminals is principally that of the diode. The V100EB diode is rated to withstand a maximum bias of 10 Vdc, and at this voltage the reverse leakage current is only 100 nanoamperes.

The ac voltage presented to the capacitance terminals must be small, with respect to the dc control voltage; otherwise the ac voltage, instead of the dc, will determine the diode capacitance. In most instances, this will occasion little difficulty, since the signal voltage will ordinarily be small, compared with the 1 to 12 volts for control voltage.

Several VCDs may be connected in parallel to increase the maximum obtainable capacitance. It must be remembered however, that the parallel connection will also increase the minimum capacitance. Thus, five type V100EB units connected in parallel in the circuit of Fig. 8-3A will provide a maximum capacitance (zero Vdc) of approximately 1300 pF, and a minimum capacitance (12 Vdc) of approximately 325 pF.

In short, all of the capacitance values in Fig. 8-3B will be multiplied by 5. At the same time, the leakage current will be ⅕ of the rated value for a single VCD (in this case 20 nanoamperes).

The electronic-type variable capacitor is smaller than the tiniest manually variable capacitor of the same capacitance. Obviously, if a potentiometer is used to vary the capacitance-controlling dc voltage, this potentiometer may well be as large as—or even larger than—an equivalent plate-type capacitor; however, this may be when the electronic capacitor is to be remotely controlled. In many applications, the variable dc voltage will already be available in a circuit in which the electronic capacitor is used.

Nos. 72-74: VCD-Tuned LC Circuits

Figure 8-4 shows several circuits in which the voltage-variable-capacitance of a VCD is employed to tune an inductance-capacitance (LC) circuit. In these examples, a type V100EB diode is shown. The capacitance of this unit varies from 260 pF when the dc control voltage is zero, to 65 pF when the voltage is 12 (see Fig. 8-3B). Table 8-1 shows the resonant frequencies corresponding to the dc control-voltage value (in steps of 1 volt from zero to 12 volts) for five different values of inductance, L1.

Figure 8-4A is the basic circuit, which is used as is in most applications of VCDs. Here, capacitor C1 prevents short circuit of the dc control voltage by inductor L1. This capacitance is so large, with respect to that of the VCD, that the VCD rather than C1 tunes the inductor. Resistor R1 acts somewhat as an rf choke, blocking the ac signal in the tuned circuit from the dc supply.

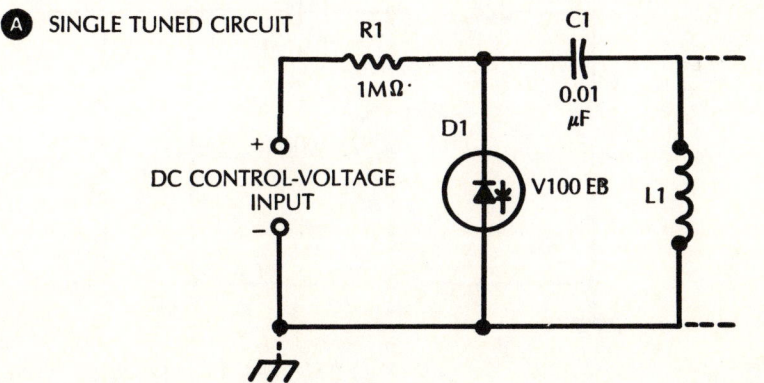

Fig. 8-4A.

Figure 8-4B shows how two circuits may be gang-tuned simultaneously by a single dc control voltage. Each of the halves in this circuit is identical with the one in Fig. 8-4A. While two circuits are shown in Fig. 8-4B, the method may be extended to as many stages as required, all being tuned by a single potentiometer or by a single available variable dc voltage. Figure 8-4C contains an additional tuning capacitor, C2. Whereas the VCD is the lone tuning capacitor in Fig. 8-4A and Fig. 8-4B, in Fig. 8-4C it is a voltage-controlled trimmer for fine-tuning the frequency in response to a dc error signal or manually adjusted dc voltage. Here, the circuit is tuned principally by means of conventional variable capacitor C2. The capacitance of the diode is effectively in parallel with C2 and therefore can retune the circuit above and below the point established by the setting of C2. In practice, the diode capacitance is chosen considerably lower than that of capacitor C2, thus insuring that C2 is the dominant frequency-determining capacitance.

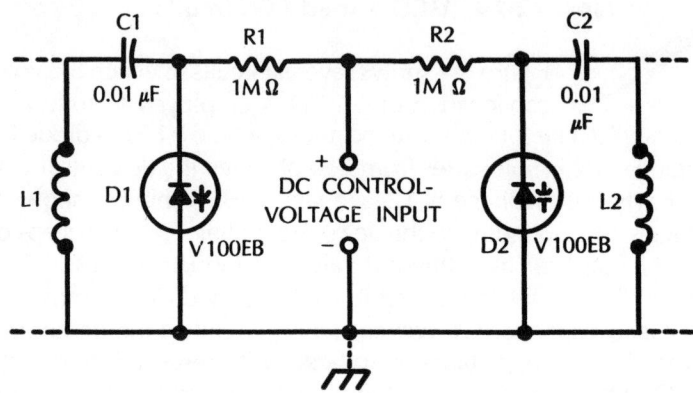

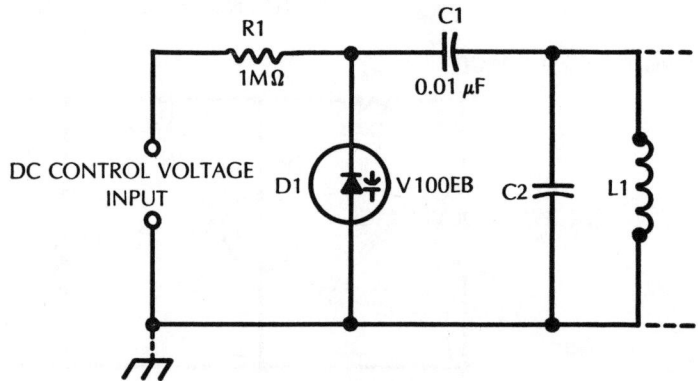

Fig. 8-4. VCD-tuned LC circuits.

The circuits in Fig. 8-4, and logical modifications of them, find use in receivers, transmitters, instruments, and control devices. When the scheme in Fig. 8-4B is extended to several stages, a small voltage-controlling potentiometer can take the place of a bulky multigang tuning capacitor. Because of the low capacitance of VCDs, these circuits are usually limited to radio frequencies. However, it is possible to parallel-connect two or more VCDs for higher capacitance.

While the V100EB diode is shown here as a practical example, other VCDs may be selected for particular applications.

DC CONTROL VOLTAGE	RESONANT FREQUENCY				
	L1 = 10 mH	L1 = 1 mH	L1 = 100 μH	L1 = 10 μH	L1 = 1 μH
0	98.7 kHz	312 kHz	987 kHz	3.12 MHz	9.87 MHz
1	124 kHz	392 kHz	1.24 MHz	3.92 MHz	12.4 MHz
2	139 kHz	441 kHz	1.39 MHz	4.41 MHz	13.9 MHz
3	148 kHz	469 kHz	1.48 MHz	4.69 MHz	14.8 MHz
4	159 kHz	503 kHz	1.59 MHz	5.03 MHz	15.9 MHz
5	166 kHz	525 kHz	1.66 MHz	5.25 MHz	16.6 MHz
6	173 kHz	546 kHz	1.73 MHz	5.46 MHz	17.3 MHz
7	178 kHz	563 kHz	1.78 MHz	5.63 MHz	17.8 MHz
8	179 kHz	566 kHz	1.79 MHz	5.66 MHz	17.9 MHz
9	187 kHz	593 kHz	1.87 MHz	5.93 MHz	18.7 MHz
10	192 kHz	606 kHz	1.92 MHz	6.06 MHz	19.2 MHz
11	193 kHz	610 kHz	1.93 MHz	6.10 MHz	19.3 MHz
12	197 kHz	624 kHz	1.97 MHz	6.24 MHz	19.7 MHz

Table 8-1. VCD-Tuned Circuit Data.

No. 75: Remotely Controlled Tuned Circuit

One of the obvious applications of the dc tuning of an LC circuit is the tuning of a distant device, such as a receiver or instrument. This technique has been used, for example, to tune a field-strength meter located out of the near zone of an antenna under test. The output current from the instrument was sent back by cable to the test shack to operate an indicating meter . It has been used also to tune a distant transmitter.

Figure 8-5 shows an example of this application. In this arrangement, the dc control voltage is transmitted through a shielded cable to the voltage-variable capacitor (D1) at the remote location.

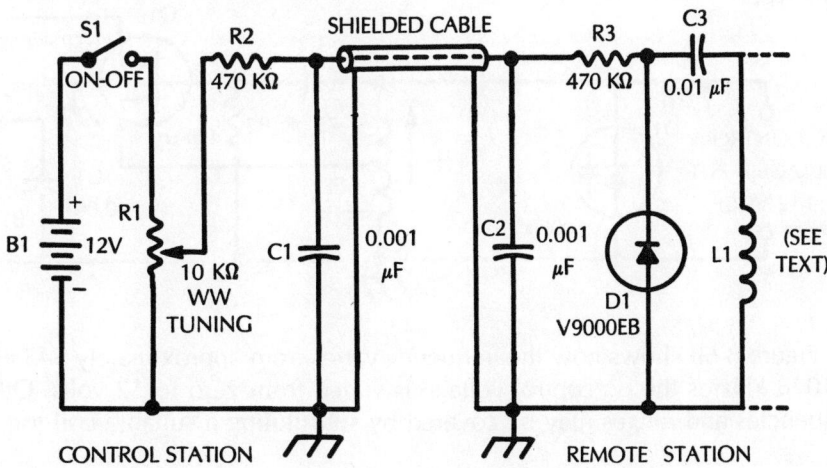

Fig. 8-5.

This voltage is adjustable from zero to 12 volts by means of a 10,000-ohm wirewound potentiometer (R1). The V900EB diode shown here has a capacitance of approximately 260 pF at zero Vdc and 65 pF at 12 Vdc, and is a low-leakage type (5 nA at 20 volts), but other VCDs may be employed, as desired.

The 470 kΩ resistors (R2 and R3) and the associated 0.001 μF capacitors (C1 and C2) effectively block the ac signal associated with the L1 tuned circuit from the dc supply. The 0.01 μF capacitance of C3 is so large with respect to that of the diode that the latter, rather than C3, tunes inductor L1. The inductance of L1 is governed by the frequency of the device in which this remotely tuned circuit operates.

Table 8-1 lists five common inductor values and shows the frequencies corresponding to control voltages between zero and 12 volts for each of these inductances. The dial of potentiometer R1 may be graduated in kHz or MHz on the basis of an initial calibration of the circuit. Although a continuously variable dc control voltage is at work in the circuit as shown in Fig. 8-5, the control voltage may instead be applied in discrete steps for step-tuning of the resonant circuit.

No. 76: Voltage-Tuned Rf Oscillator

Figure 8-6 shows how a VCD can be used to dc voltage-tune a radio-frequency oscillator. Figure 8-6A gives the circuit, and Fig. 8-6B the tuning performance. The oscillator is a tapped-coil Hartley circuit employing a type HEP F0015 field-effect transistor (Q1), but any similar circuit can also be used. The oscillator tank coil (L1) is tuned by the type V100EB variable-capacitance diode (D1), which is connected across the coil through 0.01 μF capacitor C1. (The capacitance of C1 is so high, with respect to that of the VCD that the latter rather than C1 tunes the coil.) For the 600-1078 kHz range (see Fig. 8-6B), L1 must be a slug-adjusted 0.04-0.24 mH coil with a tap at ⅓ of its turns (J. W. Miller No. 9011, or equivalent).

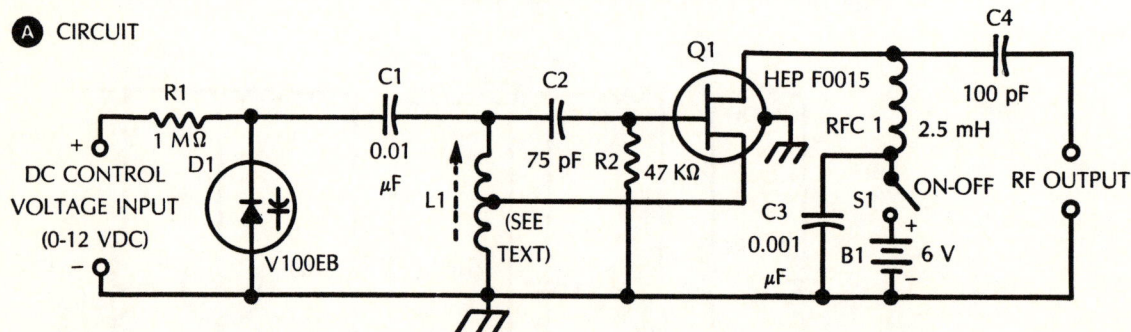

Figure 8-6B shows how the frequency varies from approximately 645 kHz to 1078 kHz as the dc control voltage is varied from zero to 12 volts. Other frequencies and ranges may be covered by substituting a suitable coil for L1.

 **PERFORMANCE**

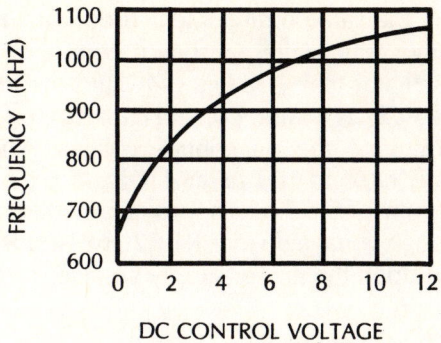

Fig. 8-6. Voltage-tuned rf oscillator.

In the initial adjustment of the oscillator, the slug of L1 is set for 645 kHz (as indicated by a suitable frequency meter connected temporarily to the rf output terminals) with the dc control voltage set to zero. Variation of the control voltage then should reproduce the curve of Fig. 8-6B, within the tolerance of individual VCDs. The oscillator draws approximately 5 mA from the 6-Vdc supply.

No. 77: Automatic FM Frequency Control

The dc output of an FM detector may be used as an error voltage to vary the capacitance of a VCD which, in turn, will automatically reset the tuning of the front end. The basic circuit for accomplishing this action is shown in Fig. 8-7.

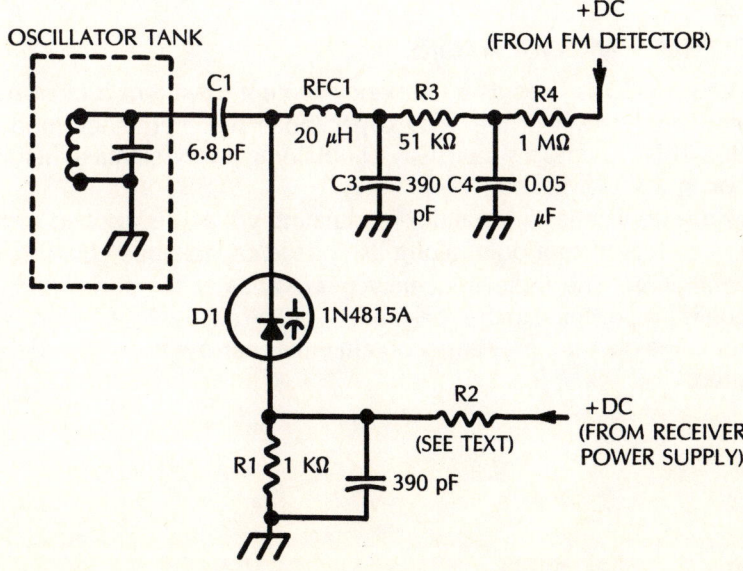

Fig. 8-7.

In principle, this is the same arrangement used with tubes or transistors for automatic frequency control (afc). As shown here, the VCD acts as a trimmer in the front-end oscillator tank (see Fig. 8-4C and the attendant discussion for a description of the VCD as a trimmer), and capacitor C1 prevents the oscillator coil from short-circuiting the dc control voltage. The 1N4815A variable-capacitance diode is rated at 100 pF at 4 Vdc.

The operating point of the diode is 5 volts provided by the dc power supply of the receiver through voltage divider R1-R2. Resistor R1 must be 1000 ohms, but R2 will depend upon the power-supply voltage, and the correct resistance may be calculated:

$$R2 = (200\ E_s) - 1000$$

where R2 is in ohms and E_s in volts. Thus, for a 35-volt supply,

$$R2 = (200 \times 35) - 1000 = 7000 - 1000 = 6000\ ohms.$$

The dc voltage must be well regulated; otherwise, the diode capacitance will fluctuate with the voltage (and so will the receiver tuning). The FM detector has no dc output, as long as the receiver is tuned sharply to a station. On detuning, however, a proportional dc voltage is delivered by the detector, and it is this voltage acting on the VCD that retunes the receiver to the station.

To compensate for the presence of the afc circuit in the oscillator, the receiver must be carefully realigned after the afc circuit has been installed. For an individual VCD, it may be necessary to adjust R2 critically around the calculated value, for optimum afc action. Since the VCD draws virtually no dc, its presence will have no effect on the FM detector. A substantial rf filter (RFC1-R3-R4-C3-C4 in the detector output line) provides essential rf isolation.

FREQUENCY MODULATORS

When a VCD is used as a frequency-varying device in a radio-frequency oscillator, modulation of the VCD capacitance will frequency modulate the oscillator. This provides a simple way of obtaining an FM signal either in a transmitter or in a signal generator.

In most applications, an audio modulating voltage is applied to the VCD which is dc biased to a point along its voltage/capacitance curve that affords linear operation. The radio frequency then varies at the audio rate, and the deviation is proportional to the peak audio-voltage swing. By adjusting the audio amplitude, the operator may easily obtain either narrow-band or wideband FM, as desired.

No. 78: VCDs Used to Frequency Modulate Self-Excited π Oscillators

Figure 8-8 shows how an audio-modulated VCD may be employed to frequency modulate a self-excited oscillator. The oscillator is a tapped-coil Hartley arrangement using a HEP S0016 bipolar transistor (Q1). For this type of oscillator, the tank coil (L1) must be tapped one-third of the way from the ground end, and such a tap is normally provided in the commercial coil (Miller No. 9012).

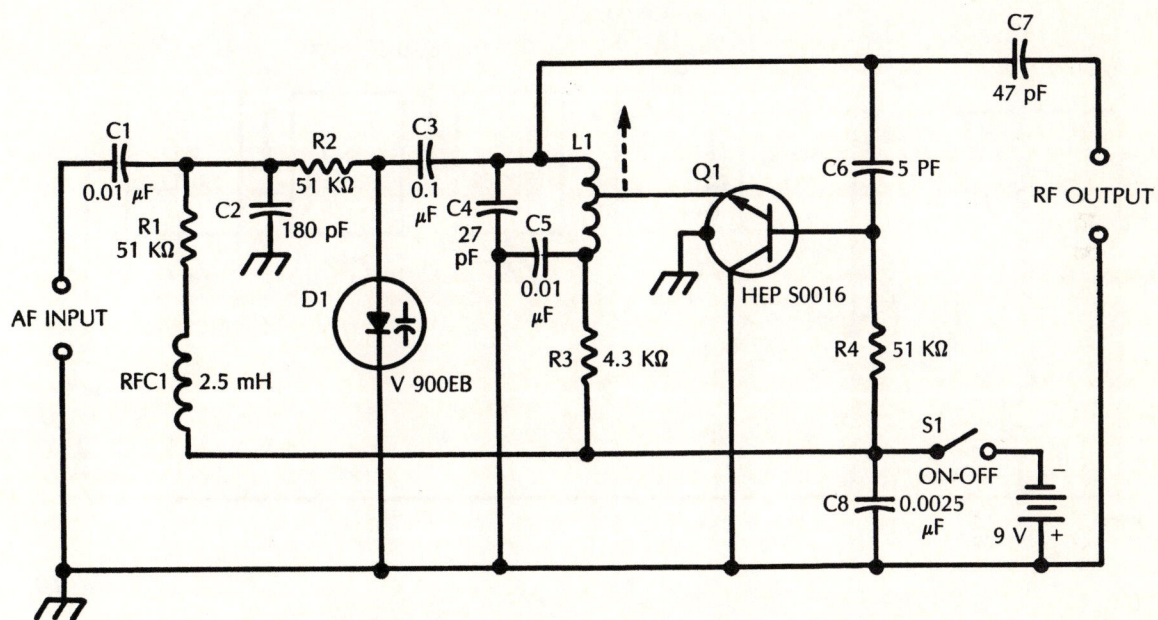

Fig. 8-8.

The oscillator is tuned principally by coil L1 and capacitor C4; the variable-capacitance diode (D1) serves as a dc-variable trimmer and is biased by the same 9-volt battery that energizes the transistor. The rf filter (RFC1-R1-R2-C2) isolates the dc supply with respect to rf, and the series capacitor (C3) prevents coil L1 from short-circuiting the dc.

The tuned circuit of the oscillator has been chosen for 1000 kHz operation, but operation may be obtained at any other carrier frequency by suitably choosing the L1 or C4 values. (The higher the carrier frequency, the greater will be the FM swing per audio volt.) For initial adjustment of the circuit, slug-tune coil L1 for the carrier frequency with zero audio volts at the af input terminals. Then apply the audio modulating voltage, and note the FM deviation for various audio amplitudes.

No. 79: VCDs Used to Fluctuate the Frequency of a Quartz Crystal

Figure 8-9 shows how an audio-modulated VCD may be used to fluctuate the frequency of a quartz crystal in an oscillator stage to frequency modulate the oscillator. The crystal frequency can be varied only a few hertz in this way, but when the oscillator is followed by several frequency-multiplier stages, this variation is multiplied along with the carrier frequency, and a useful amount of frequency modulation may be obtained in the output stage.

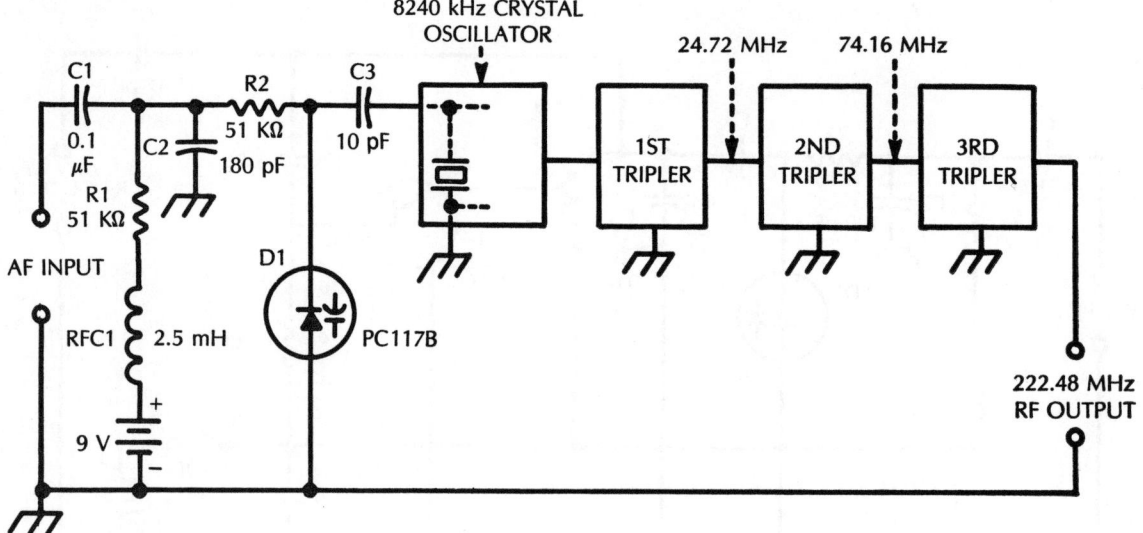

Fig. 8-9.

Thus, in Fig. 8-9, the oscillator is followed by three triplers, and here a crystal-frequency variation fluctuation of only 100 Hz would result in a fluctuation of 2.7 kHz at the final output. The circuit given here is designed to operate in the 220 MHz amateur band, the final output based upon the 8240 kHz crystal being 222.48 MHz. Varactor-type (i.e., VCD type) triplers are described in the following sections.

In this arrangement, a type PC117B, high-Q, variable-capacitance diode (D1) operates in parallel with the crystal through dc-blocking capacitor C3. Audio modulating voltage is applied to the VCD against the operating-point-setting dc bias provided by dc source B1 through the rf filter RFC1-R1-R2-C2. The resulting variation of the VCD capacitance at the audio rate fluctuates the frequency of the crystal. The deviation of the crystal frequency is proportional to the audio amplitude, and this deviation is multiplied successively by the triplers.

While the scheme shown here utilizes an 8240 kHz crystal to obtain frequency-modulated operation at 222.48 MHz, the same general arrangement may be applied for other output frequencies by suitably choosing crystal frequency and the number and type of multipliers. (The higher the final carrier frequency, the greater will be the FM swing per audio volt.)

In both the self-excited and the crystal circuit, some experimentation with the dc operating point of the VCD is necessary in order to obtain the most linear modulation. The purpose of this bias, in each circuit, is to set the operating point of the VCD to the most linear part of the diode voltage vs-capacitance characteristic.

VCD-type frequency modulators, whether employed with self-excited or crystal oscillators offer the advantages of simplicity and ease of adjustment. Moreover, the VCD requires no audio power for its operation, so the modulating source can be simply a voltage amplifier. The VCD modulator may easily be built into an existing oscillator and will occasion only the slightest readjustment of the oscillator alignment or tuning.

It is informative to note the frequencies at which FM is authorized in the ham bands. For wide-band FM: 29—29.7, 52.5—54, 144.1—148, 220—225, and 420—450 MHz. For narrow-band FM: 3775—4000 kHz, 7150—7300 kHz, 14.2—14.35, 21.25—21.45, 28.5—29.7, and 50.1—54 MHz.

FREQUENCY MULTIPLIERS

It is clear in the capacitance/voltage characteristic curve of the VCD (Fig. 8-2C) that a considerable portion of the operation of this device is nonlinear. Because of this nonlinearity, the VCD can distort an alternating current flowing through it, and accordingly can act as a harmonic generator. Indeed, the attractiveness of the VCD as a harmonic generator is now well exploited in frequency doublers and triplers.

In this application, in which the VCD is called a varactor, no dc supply is needed, so the frequency multiplier is both simple and economical. Furthermore, it is efficient (90% is typical for varactor-type doublers, compared with 50% for the tube-type), since the VCD—a reactive device—draws very little power from its driving source. Figure 8-10 shows one type of varactor doubler, and Fig. 8-11 a varactor tripler.

No. 80: Frequency Doubler

In the doubler, Fig. 8-10, the input circuit is tuned to the frequency of the driver and the output circuit to twice that frequency. In this instance, inductor L1 and capacitor C1 are tuned to 28 MHz, and inductor L2 and capacitor C2 to 56 MHz.

For these frequencies, the coils have the following specifications:

L1—7 turns No. 14 enameled wire airwound 1 in. in diameter. Space to winding length of 1 inch. Tap 2½ turns from ground end.

L2—5 turns No. 14 enameled wire airwound 1 in. in diameter. Space to winding length of 1¼ inch. Tap second turn from ground end. Other capacitor and inductor combinations may be employed for desired input and output frequencies. A 1N4386, 50-watt, varactor (D1) is used in this circuit.

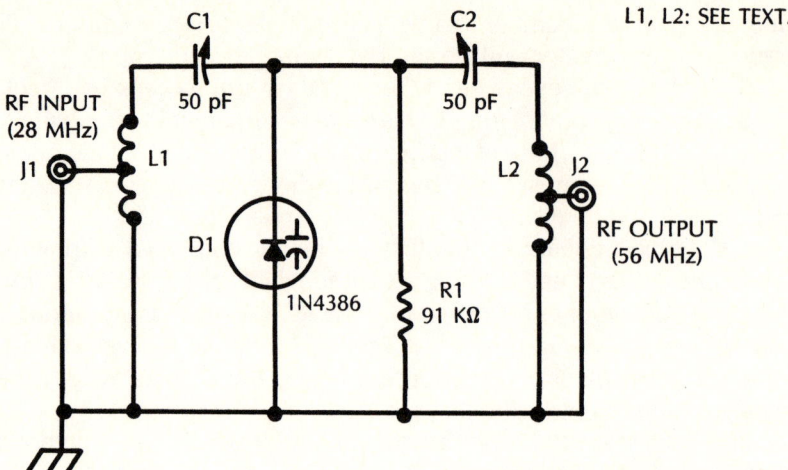

Fig. 8-10. Frequency doubler.

No. 81: Frequency Tripler

In the tripler—Fig. 8-11, the input circuit is tuned to the frequency of the driver, an idler circuit to twice that frequency, and the output circuit to three times the driver frequency. The purpose of the idler (L3-C2) is to reinforce the tripler action and increase the overall efficiency of the circuit.

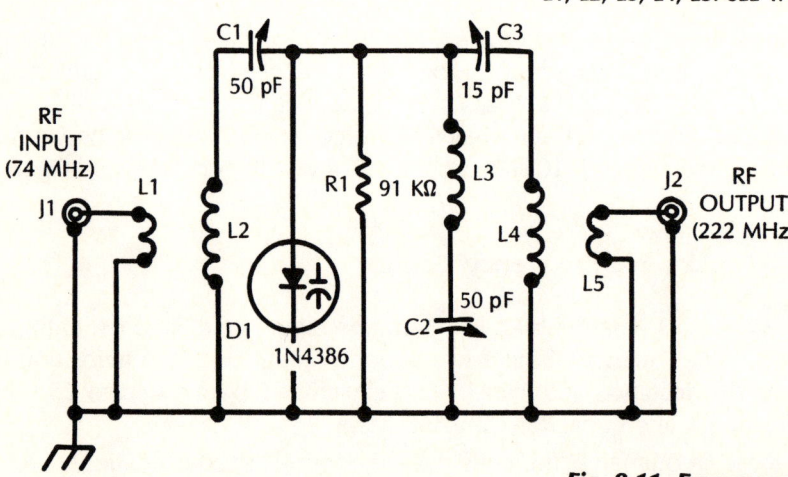

Fig. 8-11. Frequency tripler.

In this instance, inductor L2 and capacitor C1 are tuned to 74 MHz, inductor L3 and capacitor C2 to 148 MHz, and inductor L4 and capacitor C3 to 222 MHz. The coil specifications for these frequencies are given in Table 8-2. Other capacitor and inductor combinations may be employed for desired input and output frequencies. A 1N4386, 50-watt, varactor (D1) is used in this circuit.

L1.	3 Turns No. 22 Enameled Wire Closewound on Same Form as L2. Space ⅛ In. From Ground End of L2.
L2.	35 Turns No. 22 Enameled Wire Closewound on 1 In. Diameter Form.
L3.	16 Turns No. 14 Enameled Wire Airwound 1 In. In Diameter. Space to Winding Length of 1¼ Inch.
L4.	3 Turns No. 14 Enameled Wire Airwound 1 In. In Diameter. Space to Winding Length of 1 Inch.
L5.	1 Turn No. 14 Enameled Wire Solidly Mounted ⅛ In. From Ground End of L4.

Table 8-2. Coil-Winding Data.

Varactor doublers and triplers can be operated in cascade in various combinations to obtain desired output frequencies from available driver frequencies. While doublers and triplers are usual in varactor multiplier stages, frequency quadrupler action also has been reported.

No. 82: Rf Harmonic Intensifier

The nonlinearity of the VCD, so useful in frequency multipliers, can be enlisted to accentuate the harmonics of a signal source, such as a frequency standard. The very high-order harmonics from such a source are often too weak to be useful, and can be brought up to useful amplitude by means of a simple, untuned intensifier circuit.

Figure 8-12 shows the circuit of a harmonic intensifier designed primarily for use with a 100 kHz crystal oscillator used as a secondary frequency standard bandspotter.

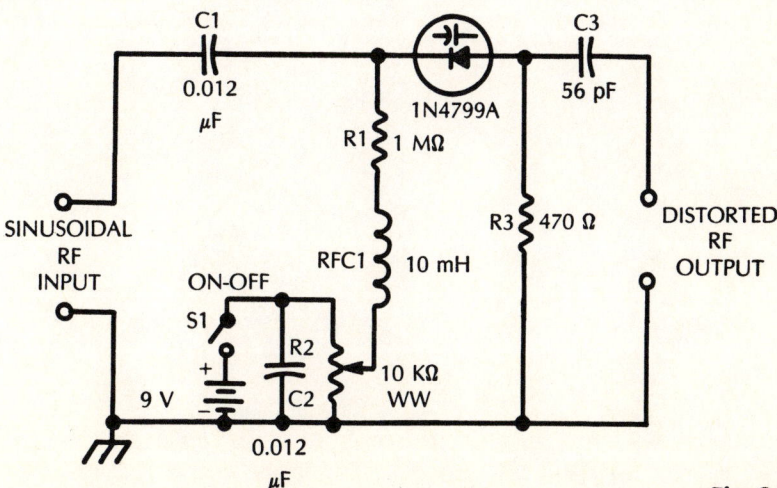

Fig. 8-12.

In this arrangement, a 1N4799A diode is biased by means of the dc circuit (B1-R2) to the most nonlinear point in the diode capacitance/voltage characteristic, as evidenced by the strongest harmonic output. The rf input-signal current flowing through the VCD and resistor R3 in series then is distorted by this nonlinear response. The output voltage, developed across resistor R3, accordingly contains a high percentage of harmonics. The intensifier circuit produces negligible loading of the signal source.

Use of the circuit is straightforward: Close switch S1 and adjust potentiometer R2 for the strongest signal, as heard in the receiver, monitor, frequency meter, or other device with which the signal source is used. The proper setting of R2 can mean, for example, hearing or not hearing the 15 MHz harmonic of a 100 kHz oscillator.

The intensifier can be built into a small case. Battery drain is 0.9 mAdc.

9
Zener Diodes

ANODE

CATHODE

Fig. 9-1.

The zener diode, also called avalanche diode or breakdown diode, is a silicon junction diode. It is specially processed to give a nondestructive breakdown (sudden large increase in current) at a specified value of reverse voltage. Around this point—termed the zener voltage, avalanche voltage, or breakdown voltage—a small change in voltage produces a large change in current, and vice versa.

This action is enlisted in numerous ways in voltage regulators, control circuits, pulse generators, threshold devices, amplitude limiters, and related applications. In some instances, the zener diode performs functions which are otherwise obtainable only with more complex devices.

Zener diodes are available in a wide range of current, breakdown voltage, and power dissipation ratings. The conventional unit (see circuit symbol in Fig. 9-1) is a single diode which is dc operated; however, special ac units, consisting of a back-to-back connected zener pair, also are available. Also, two conventional zener diodes may be connected in series back-to-back for ac service.

THEORY

Figure 9-2A shows the basic zener-diode circuit. In this arrangement, the adjustable dc source (B1) biases the diode (D1) in the reverse direction (anode negative, cathode positive), causing a reverse current (I_R) to flow through the diode and series resistor R_S. Over much of the range of the reverse voltage (V_R) the current is tiny, as shown in Fig. 9-2B, owing to the normally high reverse resistance of the silicon junction. At a critical value (V_B, the breakdown voltage), the reverse current begins to increase sharply, and for a very small increase in voltage beyond point V_B it will increase rapidly to point I_M.

The breakdown point is rounded to some extent, this curvature being referred to as the "knee." The sharper the knee, the better the diode performs in most applications. In zener diodes of some types, the knee is so sharp that no curvature is visible unless the scale is expanded sufficiently to reveal it.

In most applications, the zener diode is dc biased to a point (such as zener voltage V_z and the corresponding zener current I_z in Fig. 9-2B) somewhere in the breakdown region. The forward conduction characteristic is usually of no interest, so it is shown by a dotted line in Fig. 9-2B. With the diode operating in the zener region, a large change in current produces a small (virtually zero) change in diode voltage drop. Conversely, a small change in voltage produces a large change in current.

In applications of the zener diode, the input (applied, or signal) voltage corresponds to the adjustable-voltage battery, B1 in Fig. 9-2A, and the output voltage is chosen either as the voltage drop across the diode or the voltage drop across the resistor. For a given change in input voltage and therefore in diode current (I_R), the diode voltage drop is virtually constant because of the shape of the conduction characteristic (Fig. 9-2B), and the voltage drop across the resistor changes rapidly for the same reason. The constant diode voltage in the face of considerable input-voltage changes suits the zener circuit to constant-output applications, such as voltage regulation, peak clipping, and amplitude limiting.

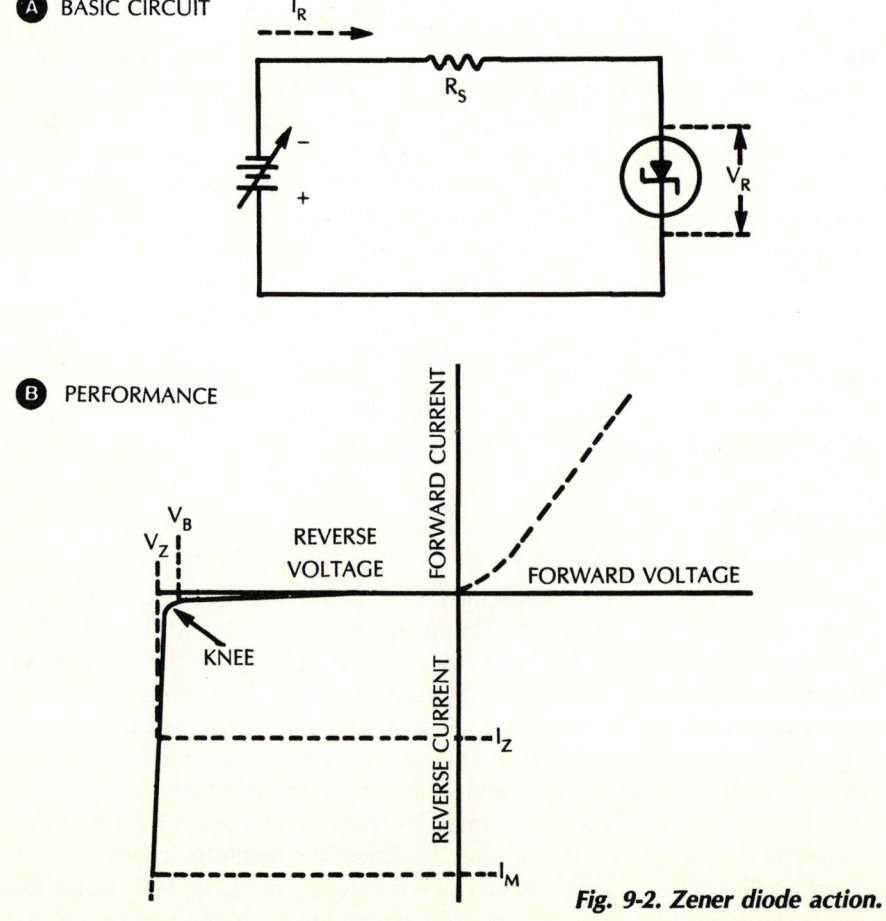

Fig. 9-2. Zener diode action.

No. 83: Simple dc Voltage Regulator

The zener diode is best used for dc voltage regulation. This application exploits the virtually constant voltage drop across a zener diode operated in series with a current-limiting resistor. As the input voltage applied to the series combination fluctuates, the resulting current through the combination fluctuates proportionately, but the output voltage taken across the diode remains virtually unchanged.

Figure 9-3 shows the circuit of the simplest dc voltage regulator employing the zener method. With the 1N4020 diode shown here, the output of the circuit is 12 volts, up to 105 milliamperes. The series resistor, R1, is a 95-ohm, 2-watt unit; and because this is not a stock resistance, it must be made up by connecting a 75-ohm and a 20-ohm unit in series (individual zeners may require some adjustment of this resistance for best voltage regulation).

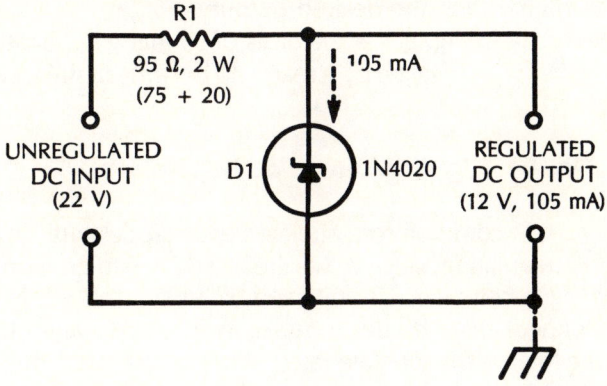

Fig. 9-3.

When the circuit is unloaded, the full current of 105 mA flows through the diode; but when the circuit is loaded, the current divides proportionately between diode and load, the total always being 105 mA. But as the load current fluctuates, the output voltage undergoes only a tiny change (see Fig. 9-2B). Similarly, when the input voltage fluctuates, the diode current correspondingly fluctuates; but large current fluctuations of this sort produce virtually no change in output (diode) voltage (again, see Fig. 9-2B). In this way, the simple circuit regulates output voltage against both input-voltage fluctuations and load fluctuations.

A simple arrangement such as that in Fig. 9-3 is often employed as a compact voltage regulator for the power supply in an electronic device. It is used also to stabilize the dc supply voltage at a single point in a circuit (as at the collector of a sensitive transistor). Occasionally, it is used in conjunction with a battery-type power supply to provide constant dc voltage as the battery ages. For an example of the convenience of the simple zener-diode regulator, see D1 in Fig. 3-8.

The output voltage is determined by the zener-diode ratings, and so is the output current. The 1N4020 unit shown in Fig. 9-3 is a 12-volt, 105 mA diode. A wide variety of ratings may be found in the diode manufacturer's literature, so that a zener diode is available for every standard output voltage and many currents. For a particular diode and an available input voltage, the series resistance (R1) must be calculated:

$$R1 = (E_i - E_o)/I$$

where E_i is the available input voltage (volts), E_o is the output voltage (diode voltage, volts), and I is the rated zener current (amperes).

Thus, for 18 volts output, a 1N3795A diode is available and is rated at 21 mA; and for this diode to be operated with a 25-volt supply, R1 = (25 − 18)/0.021 = 7/0.021 = 333 ohms. The power dissipated by this resistance would be I^2 R1 = 0.021^2/333 = 0.000441/333 = 0.0013 mW. Because of the inherent voltage drop across the series resistor in the zener circuit, the supply voltage must always be higher than the desired output voltage.

The simple circuit in Fig. 9-3 is the basis of so many applications of zener diodes that it merits the time taken by a reader to become familiar with its design and operation.

No. 84: Higher Voltage dc Regulators

It is common to associate zener diodes with low-voltage dc regulation, since it was the first device to provide simple regulation of dc voltages between 1.8 volts and 50 volts. And while it is entirely practicable to connect zener diodes in series for higher voltages (the regulated voltage being the total of the separate rated zener voltages) so long as the diode current ratings are identical, there are available single zener diodes that will do the job by themselves. Thus, in some applications where current requirements are the same for both devices, a single higher voltage zener diode can replace a gaseous regulator tube such as OA3/VR75 (75 volts), OA2/VR150 (150 volts), OB2/VR105 (105 volts), or OB3/VR90 (90 volts).

Figure 9-4 shows six dc voltage regulators of the simple type, each employing a single zener diode.

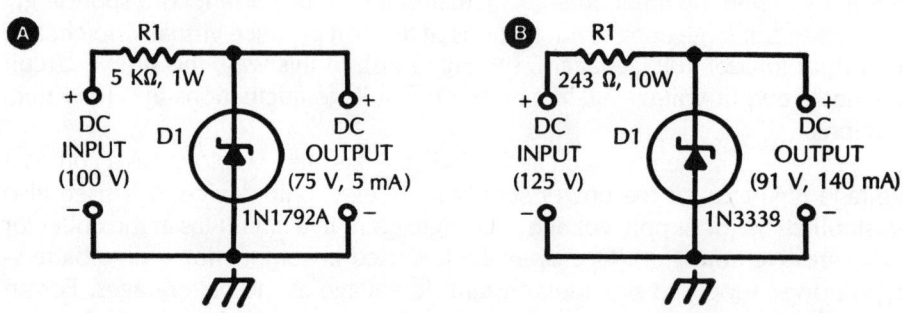

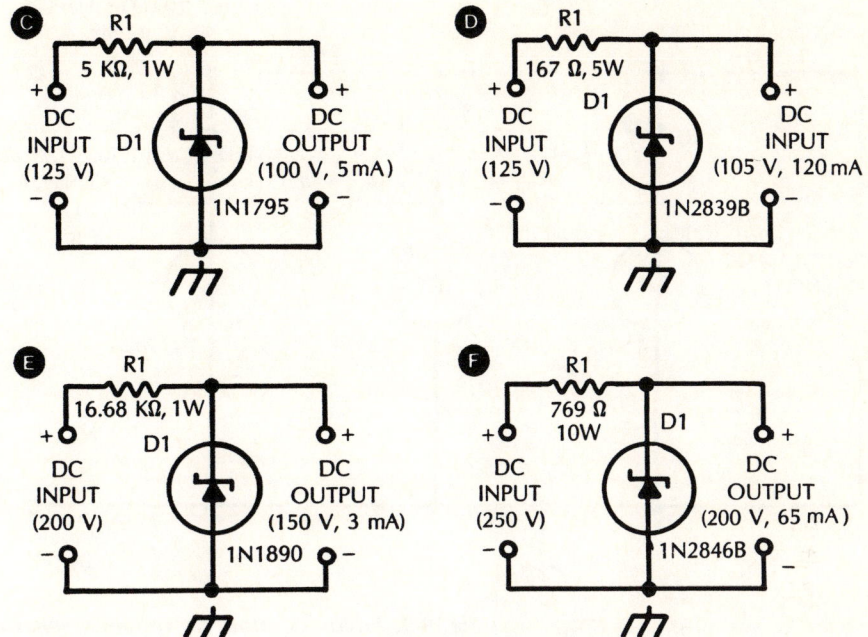

Fig. 9-4. Simple higher-voltage dc regulators.

Except for the use of a higher voltage diode, each of these circuits is identical with the basic regulator circuit discussed in the preceding section, and that section should be studied for familiarity with the circuits. In each instance, the series resistor (R1) has been worked out for a typical input voltage; but since most of these resistances are not stock values, they must be obtained by series connecting lower values.

Electronic equipment of various types—including control devices, communications systems, test instruments, and data processing devices—can utilize simple dc voltage regulators of the zener type, such as are depicted by Fig. 9-4.

No. 85: Multiple-Output dc Voltage Regulator

It is often convenient to obtain several regulated output voltages from a single supply voltage. In the parallel circuit shown in Fig. 9-5, three such outputs (10, 18, and 30 volts) are obtained from a single 48-volt unregulated supply. While three outputs are shown in this arrangement, the same scheme may be used for as many more separate voltages as required, as long as the unregulated supply can provide the total required current. An examination of the circuit shows that each leg is a simple regulator of the type described previously.

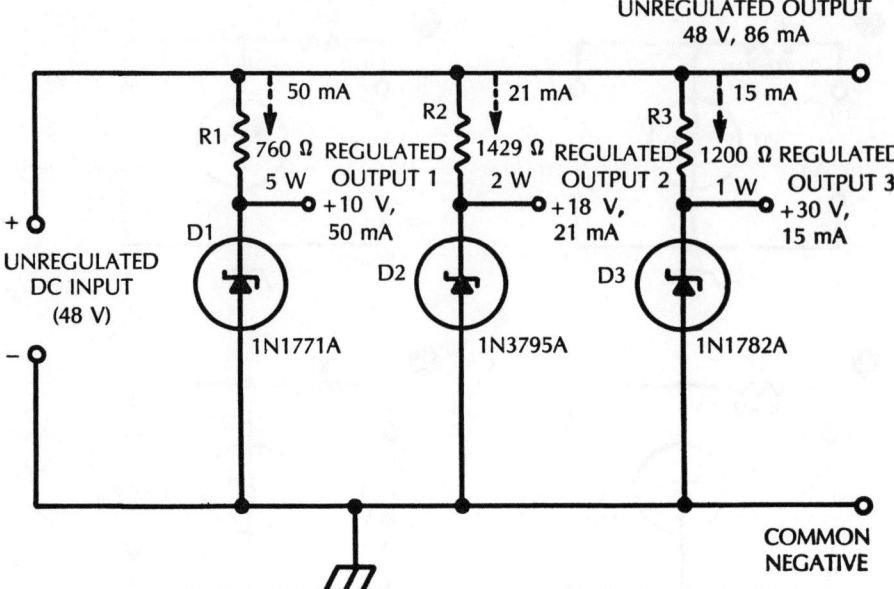

Fig. 9-5. Multiple-output dc voltage regulator.

In Fig. 9-5, the 1N1771A diode (D1) supplies 10 volts at 50 mA, the 1N3795A diode (D2) 18 volts at 21 mA, and the 1N1782A diode (D3) 30 volts at 15 mA. The series resistors (R1-R2-R3) have been worked out for the individual currents of these diodes; but since the first two of these resistances are not stock values, they must be obtained by series connecting lower values. All of the resistances may need some adjustment with individual diodes, for best voltage regulation. For a grounded-positive circuit, it is necessary only to invert the diodes and the input voltage; do not change the position of the series resistors.

No. 86: Light-Duty Regulated dc Supply

Figure 9-6 shows the circuit of a complete, power-line-operated dc power supply with a simple zener-diode voltage regulator. This light-duty unit supplies 3.5 volts at 10 mA, and is useful for operating transistors and other low-voltage, low-current devices.

In this power supply, a 6.3-volt, 1.2-amp filament-type transformer, T1 (Stancor P-8190 or equivalent) drives a bridge rectifier composed of four 1N4001 rectifier diodes (D1 to D4). The dc output of this rectifier is applied to the 1N373 zener diode (D5) through series resistor R1. (Since the 550-ohm value of this resistor is not standard, it must be obtained by parallel connecting two 1100-ohm resistors.) The regulated dc output voltage is the voltage drop across the 1N373 zener diode, a 3.5-volt, 10-mA unit.

The 2000 μF electrolytic capacitor, C1, provides filtering of the bridge output, and is aided by the regulator circuit. An added advantage of the simple zener circuit is its filtering action. Its ability to regulate voltage also causes it to reduce ripple in the rectifier output.

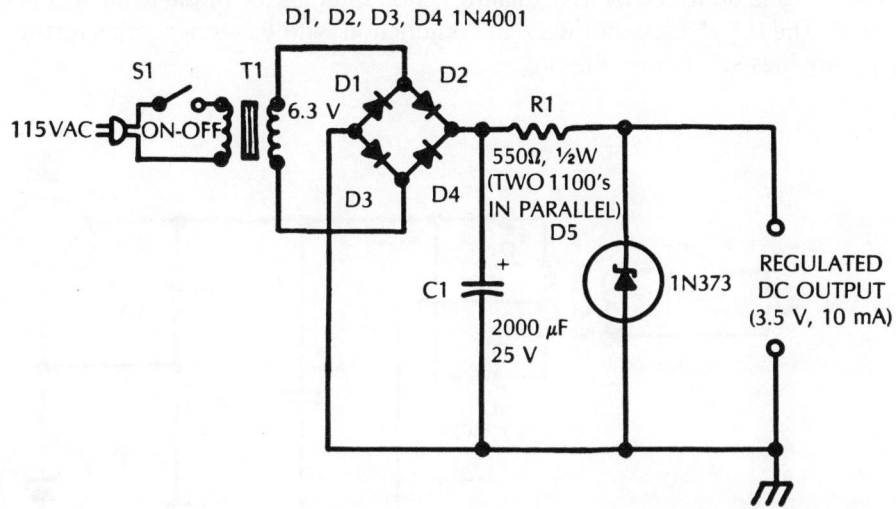

Fig. 9-6. Light-duty regulated dc supply.

Because of the small size of all of the components in this power supply, the latter may be built into other equipment in which space is limited. A typical place is a universal breadboard used for connecting circuits for transistors and integrated circuits. The unit is not limited to the output voltage and current shown in this example; higher voltage or current may be obtained in the same circuit by substituting a suitable transformer, resistor, and zener diode for those shown in Fig. 9-6.

No. 87: 5 Volt, 1.25 Amp Regulated dc Supply

The simple zener-diode voltage regulators described in the preceding sections are entirely adequate in a number of applications where they supply the constancy desired. For closer voltage regulation, however, the more complex circuit (employing a control transistor) is needed, and in this latter circuit the zener diode serves as a voltage reference device. Figures 9-7 and 9-8 show circuits of this type.

Figure 9-7 shows the circuit of a power-line-operated supply delivering 5 volts at 1.25 ampere. This circuit will be recognized as the conventional voltage regulator in which a power transistor (Q1) acts as an automatic series resistor to adjust the output voltage. The operating point of the transistor is established by the HEP Z0408 zener diode, D3. Because of the high collector current of the HEP S7002 transistor, a heat sink is required.

In this supply, power transformer T1 (Stancor P-6377 or equivalent) has a 24-volt center-tapped secondary, which drives the full-wave rectifier consisting of the HEP R0090 rectifier diodes, D1 and D2. The 250 µF electrolytic capacitor, C1, filters the dc output of the rectifier, and is aided in this function by the voltage regulator, which by its regulating action smooths the ripple in the rectifier output. The 0.1 µF capacitor (C2), in conjunction with the zener series resistor (R1) provides still further filtering.

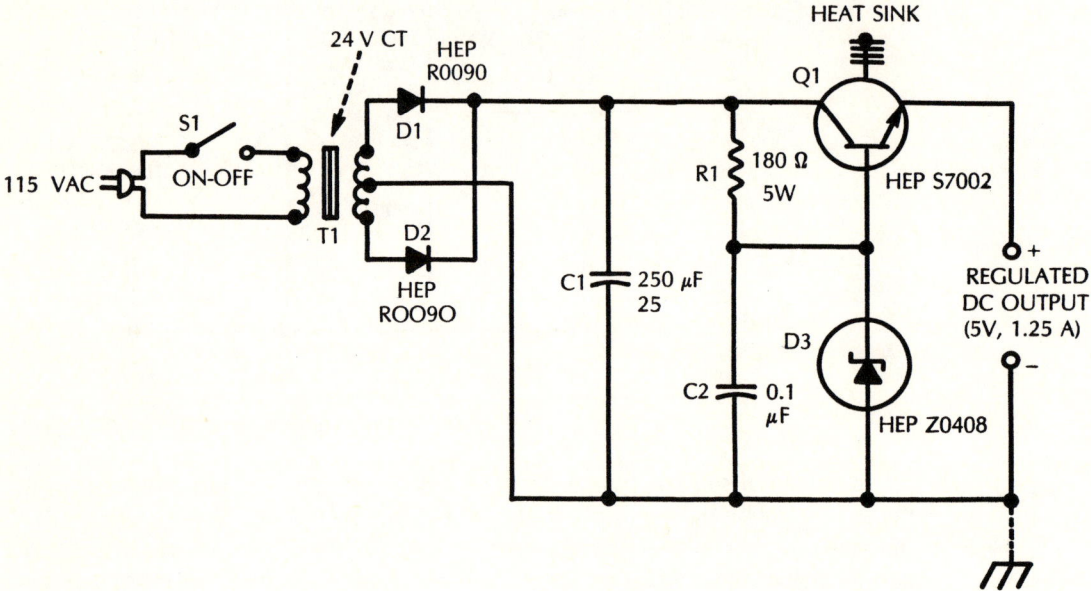

Fig. 9-7. 5 V, 1.25 A regulated dc supply.

The HEP Z0408 zener diode is a 6.2-volt, 100-mA unit. The series resistor, R1, sets the diode current to 100 mA, and its 180 ohms is a stock value; however, this resistance value may need some adjustment with an individual diode.

This power supply circuit may easily be adapted for higher voltage or higher current service by appropriately changing the transformer, rectifiers, transistor, and zener diode. The zener voltage rating of the new diode must be equal to the desired new output voltage, and the new transistor must be able to pass the new current safely.

No. 88: 18 Volt, 1 Amp Regulated dc Supply

Figure 9-8 shows the circuit of a conventional voltage-regulated power supply which is operated by an ac power line and delivers 18 volts at 1 ampere. This unit is similar to the one described in the preceding section, but employs a bridge rectifier (RECT1) instead of a full-wave, center-tapped rectifier.

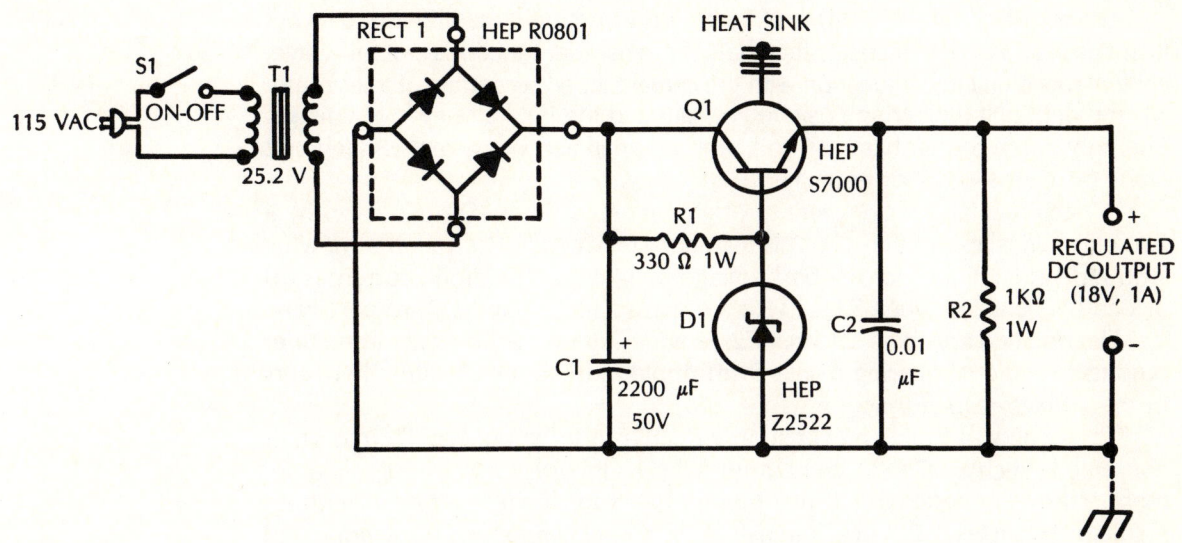

Fig. 9-8.

In this arrangement, transformer T1 (Stancor P-6469 or equivalent) delivers 25.2 volts to the bridge rectifier which is rated at 50 PRV. The dc output of the rectifier is filtered by 2200 µF electrolytic capacitor C1 and by the ripple-reducing action of the voltage regulator. Additional filtering action is provided by 0.01 µF capacitor C2.

The HEP S7000 power transistor (Q) serves as the automatic resistance element in the regulator, and the HEP Z2522 zener diode (D1) serves as the voltage reference. This diode applies a constant +18-volt potential to the base of the transistor. For correct diode operation, the 330-ohm series resistor, R1, establishes a current of 21 mA in the diode. (R1 may need some resistance adjustment with an individual diode.) Because of the high current (1 ampere) flowing through the transistor and the external load, Q1 requires a heat sink.

This solid-state power supply is entirely conventional and demands no special techniques in its construction, except that adequate ventilation should be provided for the rectifier and transistor.

No. 89: Voltage-Regulated Dual dc Supply

The power-line-operated supply shown in Fig. 9-9 provides separate +12-volt and −12-volt outputs, each at 80 mA. A unit of this type is useful for powering devices, such as integrated circuits, that require two identical positive and negative voltages. For regulation of the output voltages, two circuits of the simple zener type, each based upon a HEP Z0415 zener diode, are operated from the rectifier.

The rectifier itself (RECT 1) is of bridge construction (type HEP R0801), but is not used as a bridge. Instead, each half of the rectifier functions as a full-wave, center-tapped unit in conjunction with the center-tapped secondary of transformer T1, the right half delivering positive voltage, and the left half negative voltage. The power transformer has a 24-volt center-tapped secondary and is rated at 2 amps (Stancor P-6377 or equivalent).

In each regulator leg, current through the zener diode (D1-D2) is set by a 220-ohm series resistor (R1-R2). This resistance may need to be varied somewhat with individual diodes for the best voltage regulation. The diode current is 80 mA and is obtained when R1 and R2 are correct for D1 and D2, respectively.

Electrolytic capacitors C1 and C2 provide filtering action for the rectifier outputs. Additional filtering results from ripple reduction that naturally occurs in the voltage regulator action.

The small size of this power supply enables it to be built easily into other equipment, such as IC testers and universal breadboards for IC circuit development. The same basic circuit may be employed for higher voltage or current if suitable changes are made in transformer, rectifier, resistors, capacitors, and diodes.

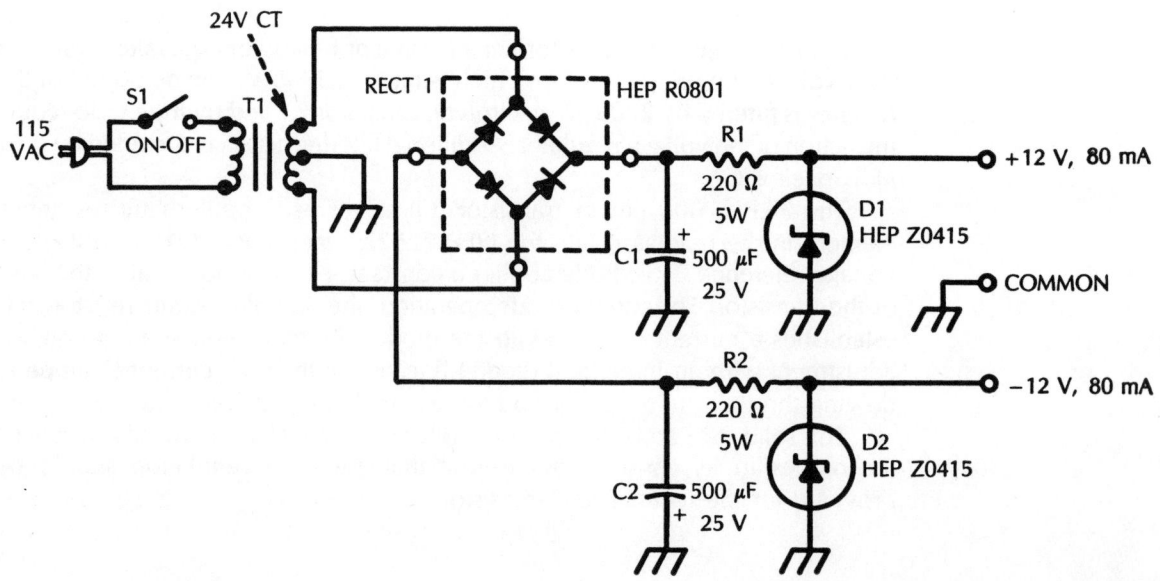

Fig. 9-9.

No. 90: Regulated Voltage Divider

Figure 9-10 shows how zener diodes may be connected in series to produce a voltage divider that gives a regulated dc voltage at each tap. For this application, each of the diodes must have the same current rating, so that the current through the string will be a single value determined by series resistor R1.

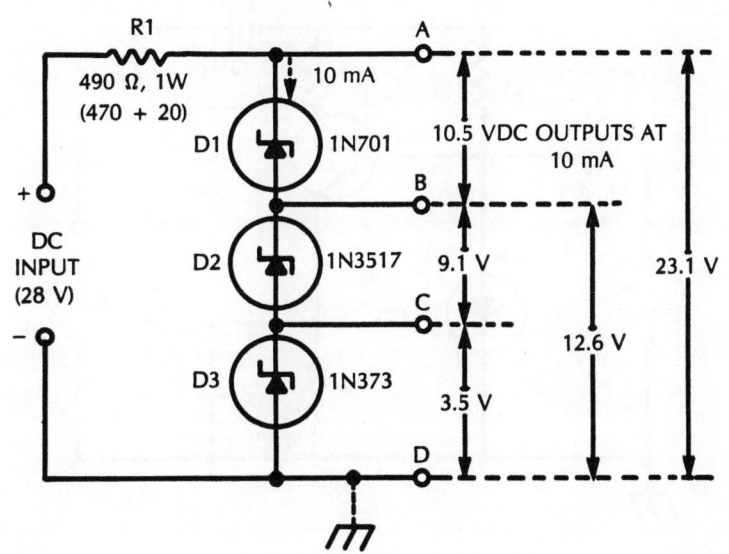

Fig. 9-10.

In the divider shown here, regulated output voltages of 3.5 volts, 9.1 volts, 10.5 volts, 12.6 volts, and 23.1 volts are provided, all at 10 mA. Any other desired combination of voltages may be obtained by employing zener diodes rated at those voltages, insuring that each of the diodes has the same zener current rating, and employing a series resistance (R) that will limit the current in the string to that value for the available input (supply) voltage. Note that three of the voltages (3.5 volts, 12.6 volts, and 23.1 volts) are with respect to the common (negative or ground) terminal, whereas two of the voltages (9.1 volts and 10.5 volts) are above ground.

The divider in Fig. 9-10 operates from a 28-volt supply; hence, the series resistance (R1) required for 10 mA is 490 ohms. This is not a stock resistance value, and must be obtained by series connecting a 470-ohm and a 20-ohm resistor. The resistance may need some adjustment with individual diodes, for best voltage regulation.

The regulated voltage divider is just as compact as one made up of resistors, since the diodes are of the same size as resistors. And it eliminates the troublesome voltage shifts that often occur when the loading of individual taps in a conventional divider is varied.

No. 91: Transistor-Bias Regulator

Stable operation of transistors demands that dc bias voltages be maintained constant. This is especially true of the base bias of a bipolar transistor. Figure 9-11 shows how a zener diode may be used in place of the lower resistor in a base-bias voltage divider to provide regulated dc voltage to the base of a transistor in an RC-coupled audio amplifier stage.

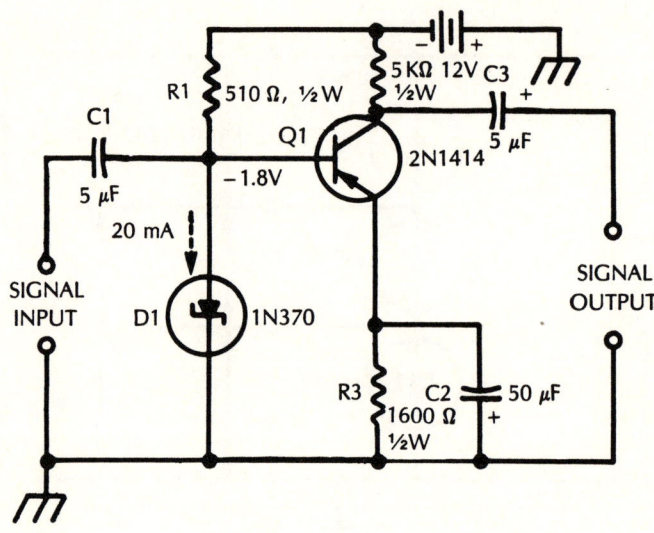

Fig. 9-11.

Here, the 1N370 diode, in conjunction with the 510-ohm series resistor (R1), supplies a constant −1.8-volt bias to the base of a 2N414 transistor. This voltage remains constant in the face of variations in supply voltage (B1) or transistor base current. Except for the substitution of the zener diode for the usual resistor, the circuit is the familiar common-emitter RC-coupled voltage amplifier.

In this arrangement, the 20 mA zener current for D1 is determined by the 510-ohm series resistor (R1) operated from the 12-Vdc supply (B1). For other bias values or supply voltages, select the zener diode that will supply that bias, and calculate the resistance needed to obtain the required zener current from the available supply voltage.

No. 92: Voltage Regulator for Tube Heater

Vacuum tubes in sensitive locations—such as variable-frequency oscillators, electrometers, Q meters, and so on—often give improved performance if their heater (filament) voltage is held constant. Figure 9-12 shows how a zener diode voltage regulator may be used to stabilize the dc heater voltage of a type 8628 high-mu triode. The heater of this tube requires 6.3 volts at 0.1 ampere. The 1N1766 zener diode supplies 6.2 volts. The slightly reduced heater voltage often results in better operation of the circuit in which the tube is used.

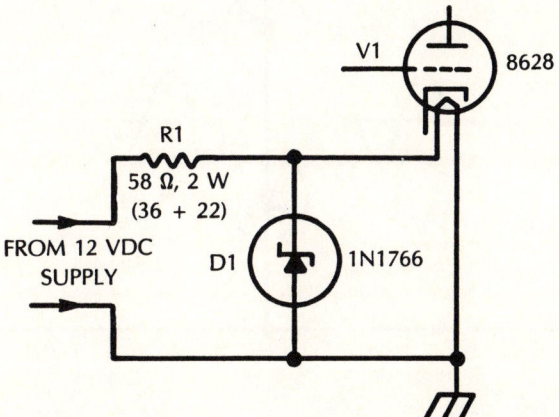

Fig. 9-12.

In this arrangement, the 100 mA current of the diode is determined by the 58-ohm series resistor (R1) operated from a 12-volt supply. Since this is not a stock resistance, it must be obtained by series connecting a 36-ohm and a 22-ohm resistor. Some slight adjustment of this resistance may be needed with an individual diode for best voltage regulation. For initial adjustment, temporarily disconnect the heater, and adjust R1 (if necessary) for a diode current of 100 mA. Then, reconnect the heater.

No. 93: Simple ac Voltage Regulator

Two zener diodes may be connected in series back-to-back to regulate the two half-cycles of an ac voltage. Figure 9-13 shows such an arrangement. Operating from 115 volts, this circuit delivers 105 volts at 120 milliamperes.

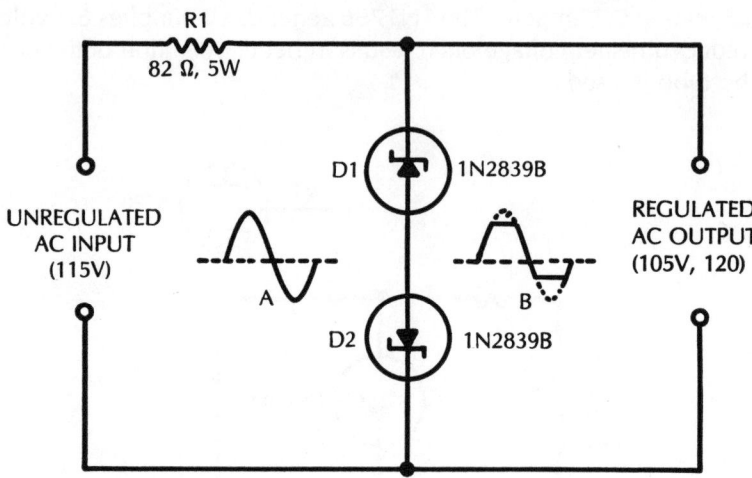

Fig. 9-13.

The ac voltage regulator is in effect two regulators, one for the positive half-cycle, the other for the negative half-cycle. When one is operating, the other is resting to all practical purposes. Thus, in Fig. 9-13 the positive regulator consists of diode D1, and the negative regulator consists of diode D2. Both diodes are served by the one series resistor, R1. When D1 is limiting the positive peak, a 120 mA zener current flows through the two diodes in series; but D2 is conducting in the forward direction and accordingly offers very little resistance, so D1 is the "working" diode at this time. Subsequently, when D2 is limiting the negative peak, the 120 mA zener current flows in the opposite direction through the two diodes in series; but this time D1 is conducting in the forward direction and offers very little resistance, so D2 is the "working" diode. This action limits both the positive and the negative peak to 105 volts, resulting in the clipped output waveform shown at (B).

For best results, D1 and D2 should be a matched pair. The common series resistor (R1) is 82 ohms, a stock resistance, but this resistance may require some slight adjustment for best voltage regulation.

The 105-volt, 120 mA regulated output shown here is typical. Other values of voltage and current may be obtained from the 115-volt input and from other input voltages by suitably choosing zener diodes and the required R1 value. The 1N2839B zener diode shown here is a 12.6-watt unit.

No. 94: Automatic Volume Limiter

The basic ac voltage regulator circuit described in the preceding section may be employed to limit the amplitude of an audio signal. This application of the ac circuit is shown in Fig. 9-14; here, the maximum value the output voltage can have is 1.8 volts.

The circuit behaves in the same manner as the power-type ac voltage regulator. In this audio circuit, two 1N370 zener diodes are connected back-to-back in series, D1 conducting during the positive half-cycle and D2 during the negative half-cycle.

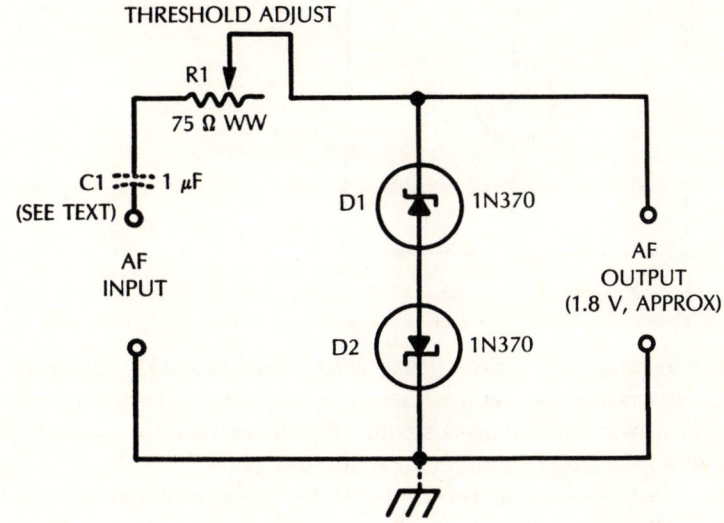

Fig. 9-14.

During the positive half-cycle, D2 behaves like a small resistance, this diode then being forward biased. Conversely, during the negative half-cycle, D1 behaves like a small resistance, this diode then being the forward-biased one.

The threshold at which the output-signal clipping begins is adjustable to some extent by means of the 75-ohm wirewound rheostat, R1. If there is direct current in the output of the signal source that drives this limiter, the zener diodes must be protected from it by insertion of a 1 µF nonpolarized capacitor, C1.

Because this is a clipper circuit, it introduces distortion in the output signal (see output waveform in Fig. 9-13). However, this distortion serves to indicate that excessive signal amplitude has occurred in the audio system and it can be corrected by needed reduction of gain, and the automatic action of the zener circuit protects the circuit and components from overdrive damage. The zener current of 20 mA is rather substantial in some audio systems, which means that the zener type of automatic limiter is most practicable in power systems. An advantage of the zener system over conventional diode clippers is its ability to operate without bias batteries.

No. 95: Dc-Equipment Protector

Because of its sharp breakdown and attendant heavy current drain, the zener diode may be used to clip a voltage transient or nullify a rise in supply voltage, thus protecting equipment from overvoltage damage. Figure 9-15 shows how an 18-volt zener diode may be connected simply in parallel with a 12-Vdc supply and the equipment (such as a receiver) that the latter powers.

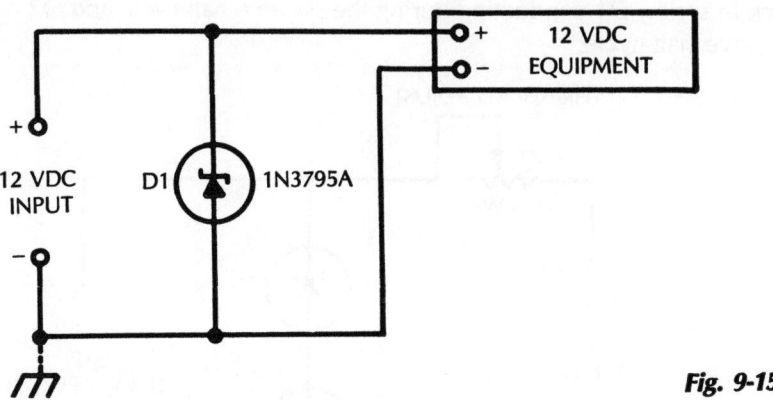

Fig. 9-15.

As long as the supply remains at 12 volts, the 1N3795A diode is in its nonconducting condition and appears as a very high resistance shunting the supply. When, however, the voltage rises to 18 volts or when a transient of that voltage value appears, D1 conducts heavily and pulls the voltage down.

The same arrangement may be employed for voltages higher or lower than 18 volts simply be selecting a zener diode having that voltage rating.

No. 96: Dc Voltage Standard—Single Stage

The constant voltage drop across a biased zener diode may be used for calibration and standardization purposes. Figure 9-16 shows two circuits, each of which provides 1.8 volts. This voltage was chosen because it is close to the voltage of a common dry cell and of a Weston standard cell. A zener voltage standard may be powered by any convenient source—e.g., a battery, powered-line-operated supply, or generator.

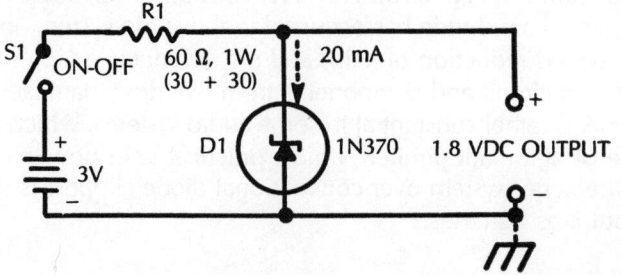

Fig. 9-16.

Figure 9-16 shows a single-diode circuit. Employing a 1N370 diode, this circuit delivers 1.8 volts when it is powered by 3 volts input. The required series resistance (for 20 mA diode current) is the nonstock value of 60 ohms which may be obtained by series connecting two 30-ohm resistors.

No. 97: Dc Voltage Standard—Two Stage

Greater constancy of output voltage, against input-voltage fluctuations, is obtained with a 2-stage circuit. The output stage then smooths out the fluctuations that normally are too low for the first stage to remove. Figure 9-17 shows such a cascade circuit. Here, the first stage, embodying a 1N1771A diode, delivers 10 volts from an 18-volt input. Any fluctuations in this 10-volt output then are removed by the second stage, which embodies a 1N370 diode. The output of this final diode is the required 1.8 volts. In this circuit, the first series resistance (R1) is a stock value; however, the second series resistance (R2) is a nonstock value and may be obtained by parallel connecting two 820-ohm resistors.

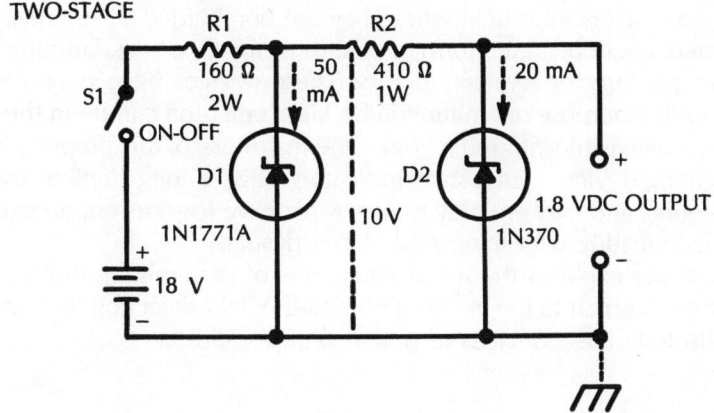

Fig. 9-17. Dc voltage standard—two stage.

Dc voltage standards may be used for calibrating such devices as meters, dc oscilloscopes, potentiometric recorders, millivolt potentiometers, thermocouple systems, and similar apparatus.

10
Emitting Diodes

Most semiconductors emit light when they are bombarded by energy. Energy bombardment could be in the form of electrons, light, or heat. Emitting diodes exploit this property of semiconductors. These devices have been designed explicitly for the purpose of emitting light. Light emission can be in the visible range or in the invisible infrared region. One major use of this property is solid-state indicating devices. Solid-state indicators have a long lifetime (typically 100,000 hours), and they are easy to use. Most have low current demand, and they interface readily with integrated circuit designs.

This chapter explores the major categories of the emitting diodes. It also covers the counterpart to the emitting diode—the light detecting transistor. The projects illustrate these devices in practical applications.

THEORY

Visible light emitting diodes are probably the most common of the emitting diodes. Previous chapters have already demonstrated the use of the light-emitting diode (LED). Combinations of LEDs into single packages as special purpose indicating devices are also common. I will discuss both seven-segment displays and bar graph displays. The infrared emitting diodes and the detecting transistors are also presented. This group of devices will also explore the use of fiberoptic cable.

Light-Emitting Diodes

A light-emitting diode converts an electrical current directly into light. By varying the type and amount of the materials used to make the pn junction, LEDs can be made to emit light at different wavelengths. Typical colors are red, green, and yellow. Simple light-emitting diodes come in three forms.

Single Color. Single color LEDs are very common, and their use has been demonstrated in many of the preceding projects. A single color LED has a pn junction on a chip of silicon. A lens covers the pn junction to focus the light that is emitted. Some LEDs have a current limiting resistor built on the chip with the pn junction. If a built-in resistor is not present, then a current limiting external resistor is needed to protect the LED.

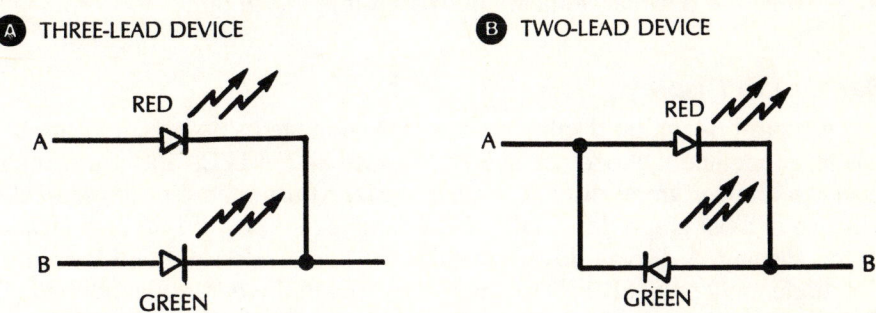

Fig. 10-1. Three-color LED.

Tricolor LED. By placing two pn junctions on the same chip that emit light at two different wavelengths, a tricolor LED can be constructed. Figure 10-1 shows two different ways that the tricolor LEDs are made.

Three-Lead Tricolor LED

In Fig. 10-1A, the LED has three leads. One lead is common to the anode of both LEDs. One lead connects to the cathode of the red LED and the other lead connects to the cathode of the green LED. If the common lead is tied to ground, then a voltage on the red or the green lead will make the LED light up. If a voltage is placed on both cathodes at the same time, then both LEDs emit light together. The mixing of the red and green light produces a yellow color.

Two-Lead Tricolor LED

Figure 10-1B illustrates a two-lead tricolor LED. Here two pn junctions are connected in opposite directions. The color of the light that the LED will emit is determined by the polarity of the voltage on the two LEDs. A signal that changes polarity will cause both LEDs to light up and produce a yellow color.

Blinking. In a number of the preceding projects, the net result of the circuit was to blink an LED on and off. LEDs are available that have a pn junction that will emit light and an IC oscillator to blink the LED. Figure 10-2 shows the internal arrangement of such a circuit.

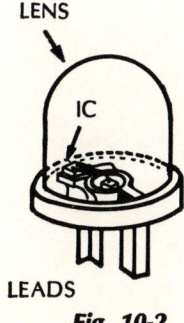

Fig. 10-2.

Seven-Segment Displays

Seven-segment displays have seven LEDs arranged in the shape of an 8. By turning on different combinations of segments, the decimal numbers from 0 to 9 can be formed. Other combinations can produce some of the letters of the alphabet. The use of a seven-segment display driver provides a very simple way to generate the decimal numbers from a 4-bit binary input.

Bar Graph Display

A typical bar graph display has ten LEDs arranged side-by-side. Each LED has its anode and cathode accessible. Because all ten LEDs are in a compact package, the bar graph display is much easier to use than ten discrete LEDs. The use of a bar graph display driver gives increased power in the application of this display. A display driver typically has two modes: dot and bar. In the dot mode, only one LED is lit at any given time. In the bar mode, all the LEDs from the one giving the high indication and below are lighted. Display drivers are also available for both logarithmic and linear displays.

Infrared-Emitting Diode Transmissions

A pn junction can be constructed that emits light in the infrared region of the electromagnetic spectrum. Sometimes these LEDs have a high-quality lens to focus the infrared light into a very tight beam. A tight beam is needed if the IR diode is going to send its light across a long distance to an infrared photodetector. The automatically opening doors that are becoming so prevalent in grocery stores use an IR emitter and detector. Information can be transmitted over considerable distances by using a well-focused lens system. The use of an IR link can be used to accomplish complete electrical isolation between circuits. In this case, the distance is usually small.

Infrared Emitting Diode Detection

Detection of light is done by the reverse process that generated the light in the first place. By bombarding the base region of a transistor with light, a bias current is produced that forward biases the phototransistor. In this way, light transmitted by an LED can be converted back into an electrical current like the one that generated the light originally.

distances. The use of fiberoptic cable also removes the line of sight restriction of light beam transmission. Fiberoptics is an area of intense research and development. The first transatlantic fiberoptic communications cable is well under way. The use of fiberoptics in telephone communications is a major application. It is also finding its way slowly into consumer items. At least one model compact disk player has a fiberoptic link between two of the circuits to obtain electrical isolation.

No. 98: Blinking LED

The CQX21 Blinking LED has an IC oscillator inside the lens. This IC pulses the LED on and off at about 3 Hz with a 5 V supply voltage. The LED will operate with a supply voltage from 4.75 V to 7 V. The blinking frequency increases with the voltage at about .2 Hz per volt. With a supply of less than 4.75 V, the oscillator will not be able to trigger the LED. As a result it will not blink.

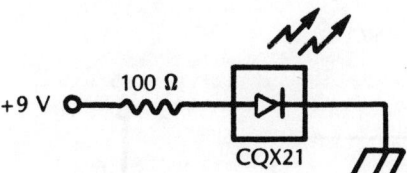

Fig. 10-3.

The circuit shown in Fig.10-3 uses a resistor, a 9-V battery and the CQX21 blinking LED. That is all that is required. The circuit can be used as an indicator on a control panel or as a marking light for a light switch in the dark. Compare the simplicity of this circuit with the one in Fig. 4-17. This is another illustration of the value of integration.

No. 99: Tricolor Logic Probe

The circuit in Fig. 10-4 is about the simplest logic probe that can be designed. It will indicate a high, a low, or a pulse with only a resistor, one inverter from a 4049 hex inverter, and a tri-color LED.

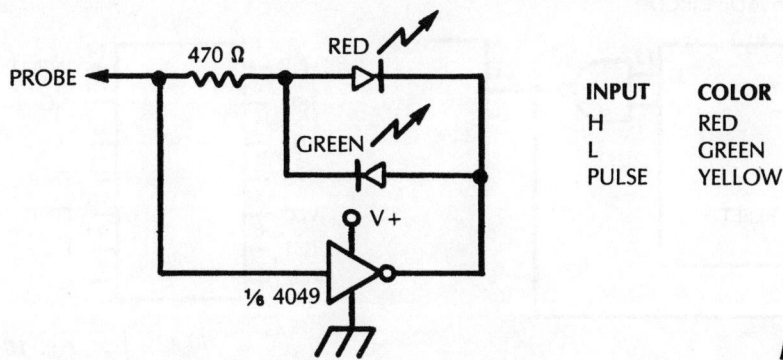

INPUT	COLOR
H	RED
L	GREEN
PULSE	YELLOW

Fig. 10-4.

The unused inputs of the hex inverter should be tied to ground. By using a CMOS inverter, the probe can be used for both TTL and CMOS circuit testing. The circuit is powered from the circuit under test. If the probe is only going to be used for TTL, a 7404 hex inverter can be used instead of the 4049.

A high input to the probe will produce a red indication if the long lead of the LED is tied to the probe tip. The green LED inside the lens will glow on a low input. A pulse on the probe will cause both LEDs to light alternately, producing a yellow indication. If the pulse frequency is low, a red and green indication will alternately flash.

No. 100: Decimal Counter

A seven-segment display and a 7448 decoder driver can be used to provide a decimal indication from a BCD count. The pinout for the 7448 is shown in Fig. 5-13. In Fig. 10-5, the BCD output of the 7490 is decoded by the 7448 for a decimal readout on the display.

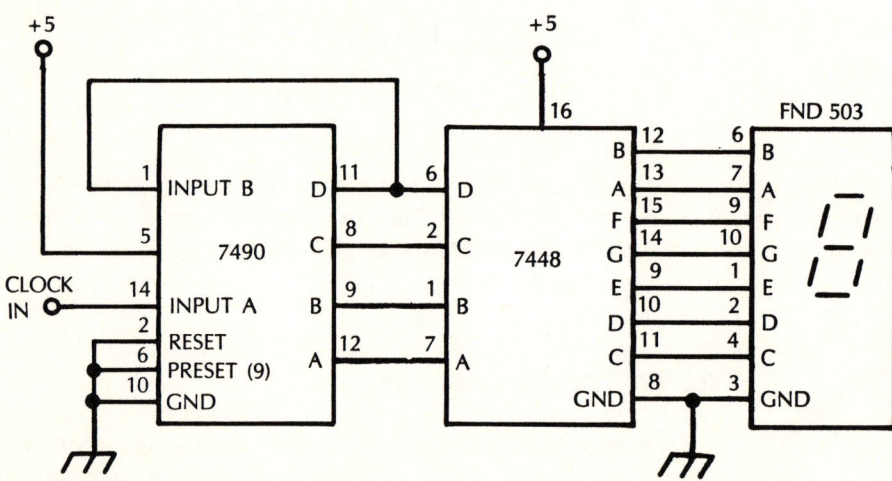

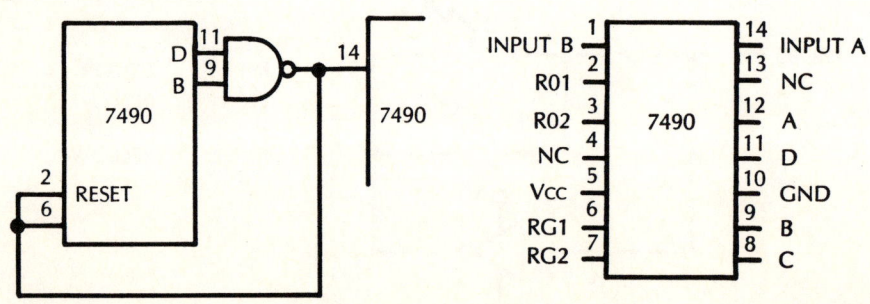

Fig. 10-5.

Such a decoding circuit simplifies reading a binary display. By using a 7447 decoder, a common anode display can be used. The FND 503 seven-segment display is a common cathode display.

BCD COUNT	DECIMAL INDICATION
0000	0
0001	1
0010	2
0011	3
0100	4
0101	5
0110	6
0111	7
1000	8
1001	9

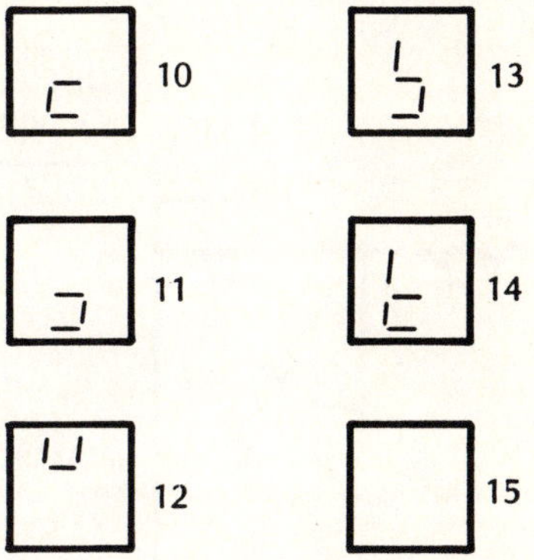

Fig. 10-6. Decoded display for numbers 10 through 15.

Any number of these circuits can be cascaded together for more digits by using a 7492 divide by 12 counter. Outputs B and D should be NANDed together to clock the input for the next stage. Each counter drives its own 7448. In this way, the ten count of the first display will trigger the next stage to indicate a 1. The NAND gate also resets the previous stage to zero. The 7493 divide by 16 counter can be used to provide a BCD output for a hexadecimal display. The seven-segment display uses the normal decimal numbers to show 0 to 9. Ten through 15 use the segments shown in Fig. 10-6 to identify the hex output.

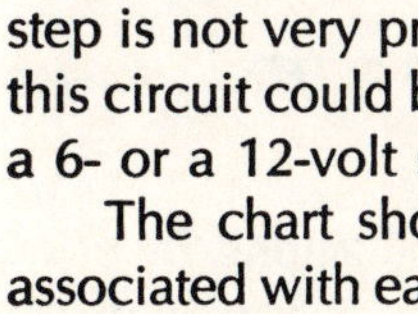

No. 101: Simple Voltmeter

A bar graph display and an LM3914 display driver can be used to construct a simple voltmeter. The 3914 is basically an analog-to-digital converter. The choice of the resistor network values will determine the voltage range of the visual indication. The circuit in Fig. 10-7 provides a visual indication of the voltage level from 2 to 13 volts. The resolution of each step is not very precise but, for some applications, it will suffice. For instance, this circuit could be used to monitor the battery voltage in a car or truck. Either a 6- or a 12-volt system can be monitored.

The chart shown in Fig. 10-7 gives a general idea of the voltage level associated with each of the 10 steps. Note the steps are not perfectly linear. The voltage range can be calculated with the following formula:

$$V_{MAX} = 1.25(1 + (R1/R2))$$

The display driver can be used in two different modes: bar and dot. The dot display will light only a single LED bar for any given input. In other words, if 5 volts is measured by the probe, the 4th segment will light. In the bar mode, a 5-volt level would cause segments 1, 2, 3 and 4 to light. In the bar mode, the display will look like the mercury inside a thermometer moving in a continuous column.

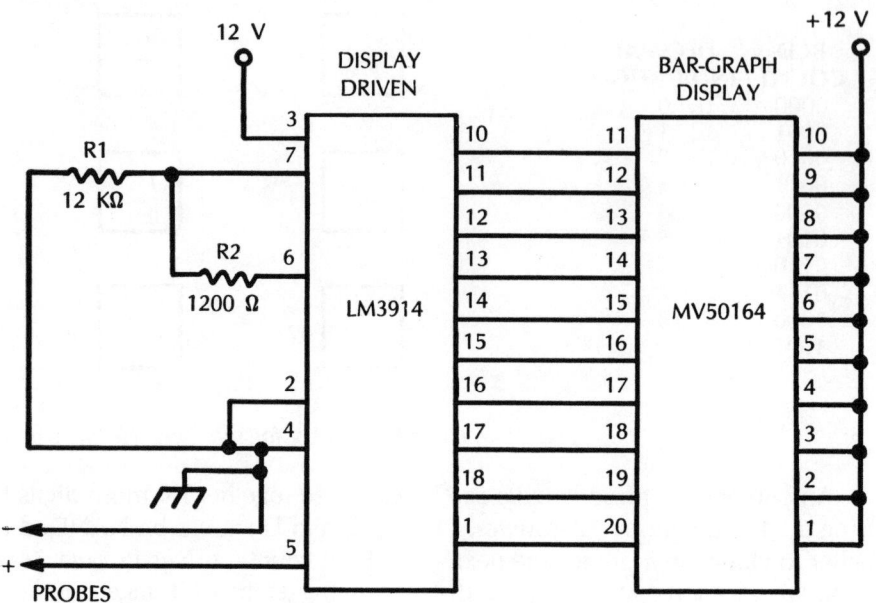

$$V_{max} = 1.25\left(1 + \frac{R1}{R2}\right)$$

LED SEGMENT	VOLTAGE
1	2
2	3.2
3	4.0
4	5.0
5	6.2
6	8.5
7	8.9
8	9.9
9	11.0
10	12.0

If pin 9 of the display driven is tied to +12 V, the display functions in the bar mode.

Fig. 10-7. LED Voltmeter.

This display configuration can be used for a number of applications. By using a thermistor, a crude thermometer can be constructed. The 3915 display driver is a logarithmic scaled analog-to-digital converter. This driver can be used to indicate audio power levels. Each level is separated by −3dB.

The 3914 and its corresponding bar graph display can be cascaded from either end to provide a large indicating range.

No. 102: Dc Controlled ac Lamp

With the aid of an optoisolator, a low dc voltage can be used to safely power a higher ac voltage. Figure 10-8 provides isolation between +9 Vdc and 120 Vac. In this case a 9-volt transistor radio battery is used to turn on a 100-W bulb. The internal triac in the MOC3010 triggers a higher voltage external triac. The lamp is controlled by S1. When S1 is opened, the lamp will go out as the ac signal passes through zero.

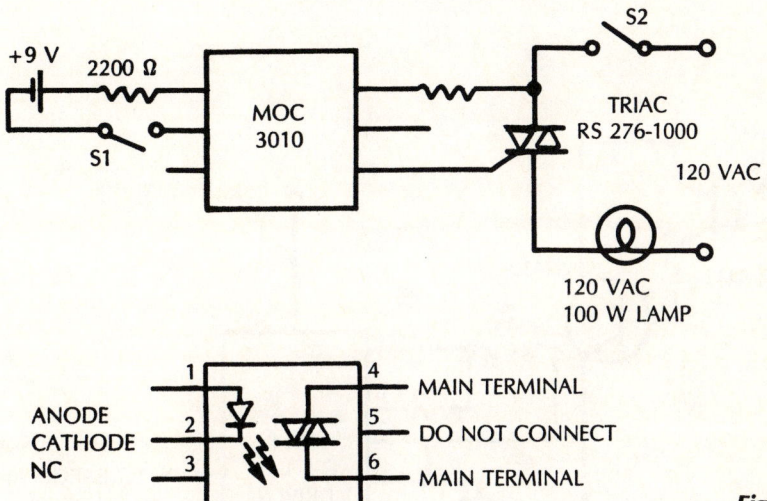

Fig. 10-8.

The optoisolator can be used for any situation requiring complete isolation between inputs and outputs. This circuit can be used to power other ac devices besides a lamp. Bells, buzzers, and motors can also be controlled.

It can also control dc devices. If a dc device is controlled, a separate switch is needed in the output-control circuitry to shut down the device. Once the triac is triggered, current will continue to flow even if the trigger signal is removed. The only way to stop the current flow is to interrupt the source of current to the triac. This can be an advantage if the optoisolator is used to control an alarm. Once the alarm is triggered, even if the trigger signal stops the alarm will continue to sound.

The triac and SCR control circuits in Chapter 12 and Chapter 13 can be controlled by the MOC3010.

No. 103: Light Meter

By using a modified version of the display driver circuit in Fig. 10-7 and a phototransistor, a simple light-level indicator can be built. Modify Fig. 10-7 by replacing R1 with a 3900 Ω resistor. Then connect the circuit shown in Fig. 10-9 to the positive input of the 3914 driver in Fig. 10-7. This new circuit will indicate the intensity of the light level that falls on the base region of the phototransistor.

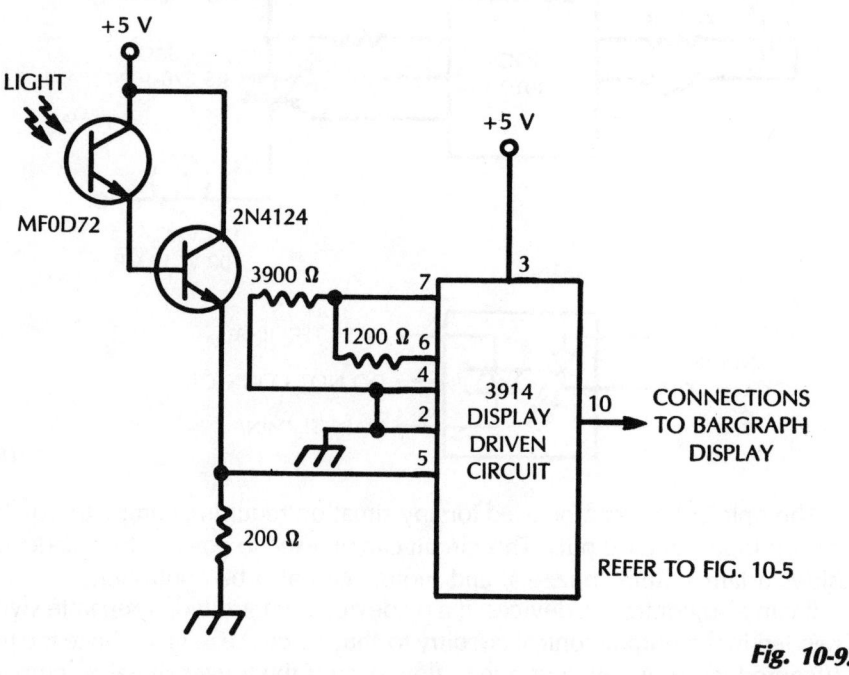

Fig. 10-9.

Although the resolution of this indicator is not very precise, it illustrates the use of the phototransistor as a sensor. This is the same kind of circuit used in cameras with a built-in solid-state light meter.

No. 104: Infrared Link

An infrared LED and a phototransistor can be used to transmit information over small distances (my setup transmitted across 6 inches without difficulty). The better the lens on the IR transmitter and receiver, the greater the distance. Just as in the optoisolator in the last section, this design approach provides complete isolation between inputs and outputs. The circuit in Fig. 10-10 can transmit digital information over a distance of several inches. The infrared LED is connected across 5 V. The phototransistor receiver picks up the infrared transmission on its exposed base region. Q1 amplifies the signal and drives the indicating LED.

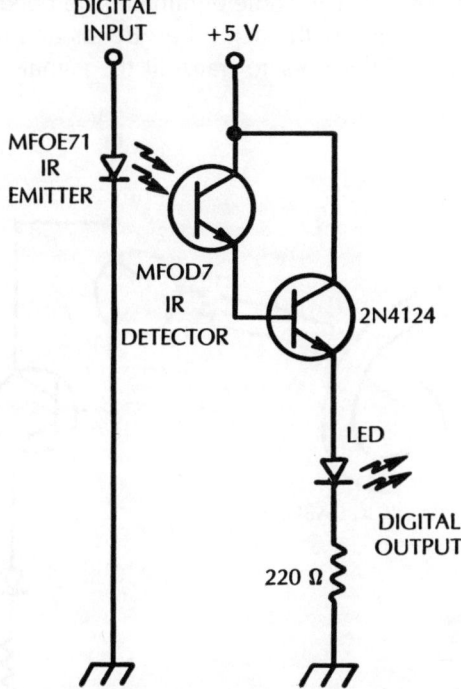

Fig. 10-10.

Instead of driving an LED, Q1 could drive the input to a counter or other digital device. The transmitter and receiver must be in line with each other for the circuit to function. The IBM PCjr personal computer had an IR data link between the keyboard and the computer console. With this link in use, the keyboard could be physically disconnected from the computer. If the distance between the receiver and transmitter becomes too great, the infrared signal will disperse too much to generate enough drive for Q1. The next project illustrates how to solve the distance and alignment problems with fiberoptic cable.

No. 105: Fiberoptic Infrared Link

Fiberoptic cable is made up of a clear-glass fiber inside a protective plastic shell. The glass fiber transmits light along the inside of the fiber without letting the light escape from the fiber. Because light cannot escape, there is very little loss of the light signal as it travels down the glass fiber. When you construct this project and compare the intensity of the LED to that of the previous project, you will see how efficient the glass fiber transmission medium is compared to using air as the medium.

I purchased 3 meters (9.8 ft.) of fiberoptic cable from Radio Shack. By simply placing the entire length of cable between the emitter/detector circuit in Fig. 10-11, the intensity of the signal more than doubled while the distance increased by more than 19 times. Impressive. The cable eliminates the need to have the emitter and detector directly in line with each other. Bends and loops do not have any affect on the ability of the fiber to transmit the signal.

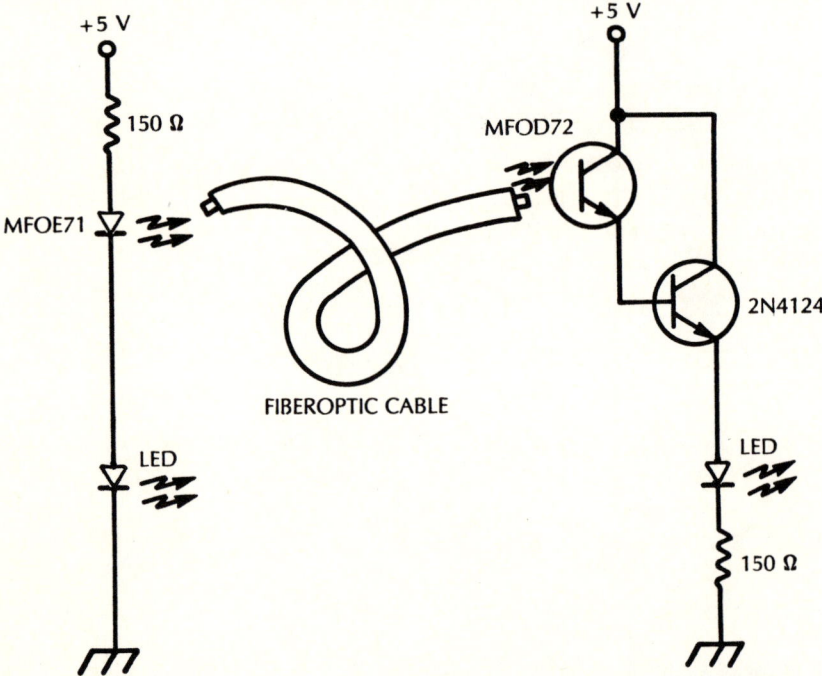

Fig. 10-11. Infrared digital link with fiberoptic cable.

11
Diac

The diac is a comparatively simple switching device of the thyristor family, in which the semiconductor contains three layers, as in a pnp transistor. Connections are made only to the two outer layers, so the diac has only two terminals and it operates on either ac or dc, conducting in either direction. Like other thyristors, the diac behaves somewhat in the manner of a thyratron tube.

The diac finds application principally in control, switching, and trigger circuits where a bidirectional, 2-terminal switching device is advantageous. It may be used by itself or in conjunction with a triac or a silicon controlled rectifier.

THEORY

Figure 11-1A shows the basic structure of the diac, and Fig. 11-1B shows the circuit symbol. In this device, the doping concentration, unlike that of the junction transistor, is the same for both junctions, and this enables symmetrical operation. Neither terminal is exclusively anode or cathode.

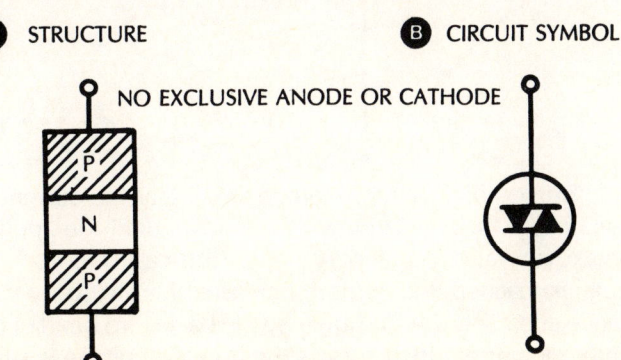

Fig. 11-1. Details of diac.

Because of the series arrangement of p and n layers inside the device, the diac can never conduct in the forward direction, but always behaves as a reverse-biased avalanche diode, regardless of the direction of an applied voltage.

Figure 11-2 illustrates the behavior of the diac. When a voltage is applied to the device, as in Fig. 11-2A, a very small leakage current flows; this is the off state of the diac. As the voltage is increased, a critical value termed the breakover voltage eventually is reached ($V_{BO}{}^+$ when the voltage is positive, $V_{BO}{}^-$ when it is negative) and at this point avalanche breakdown occurs and a heavy current suddenly flows; this is the on state of the diac. Once the diac is thus switched on by a positive or a negative voltage, the device will continue to conduct current until the voltage is removed or reduced to zero.

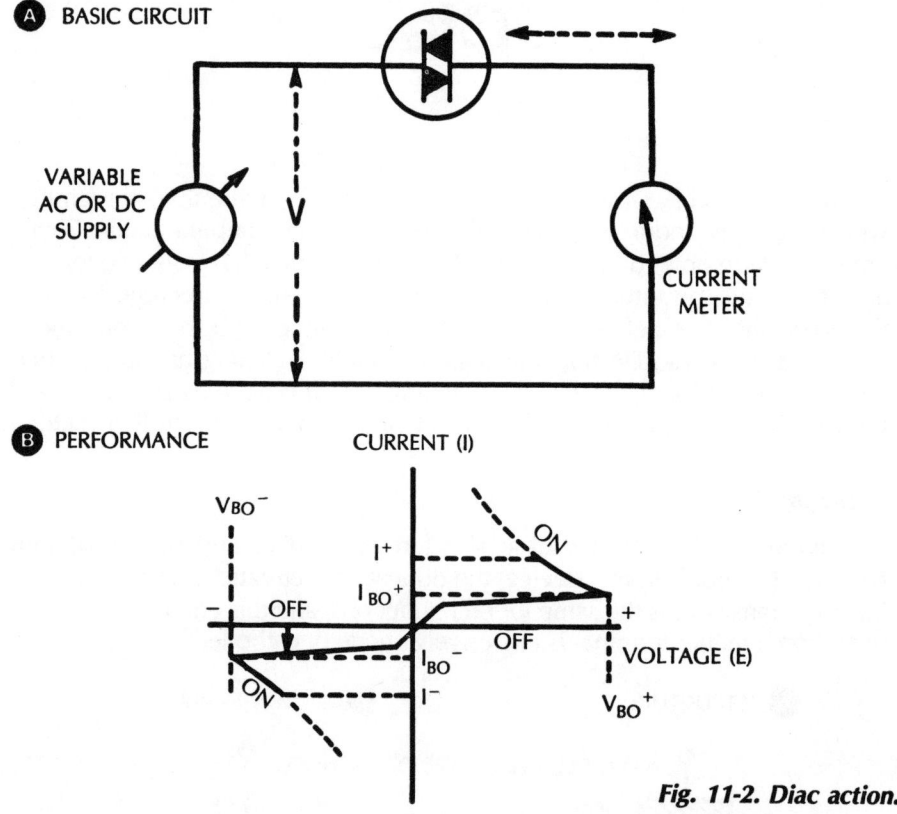

Fig. 11-2. Diac action.

Figure 11-2B depicts diac action. Here, a small leakage current ($I_{BO}{}^+$ for a positive voltage, or $I_{BO}{}^-$ for a negative voltage) flows until the applied voltage reaches the breakover value ($V_{BO}{}^+$ or $V_{BO}{}^-$, as the case may be). When the breakover voltage is reached, the current increases sharply, as from I^+ or I^-. A negative-resistance effect sets in—as shown by the backward bend of the curve—and consequently the current increases as the applied voltage is subsequently decreased.

A primary use of the diac is to supply a trigger pulse to a triac, and this application is illustrated in a number of ways in Chapter 12. However, the triggered response of the diac itself and the bilateral conduction of this device also suit it to certain applications other than triac operation. Several of these are described in this chapter and possibly will suggest still others to the experimenter.

No. 106: Amplitude-Sensitive Switch

The simplest application of the diac by itself is automatic switching. A diac appears to either ac or dc as a high resistance (almost an open circuit) until the impressed voltage reaches the critical V_{BO} value. The diac conducts when this value is reached or exceeded. Thus, this simple 2-terminal device may be switched on simply by raising the amplitude of the applied control voltage, and it will continue to conduct until the voltage has been reduced to zero.

Figure 11-3 shows a simple amplitude-sensitive switch circuit employing a 1N5411 diac. An applied voltage of 35 volts dc or peak ac will switch the diac into conduction, whereupon it will pass a current of approximately 14 mA through the output resistor, R2. Individual diacs may switch on at voltages lower than 35 volts. With 14 mA on current, the output voltage developed across the 1000-ohm resistor will be 14 volts. If the voltage source has an internal conductive path in its output circuit, resistor R1 may be omitted.

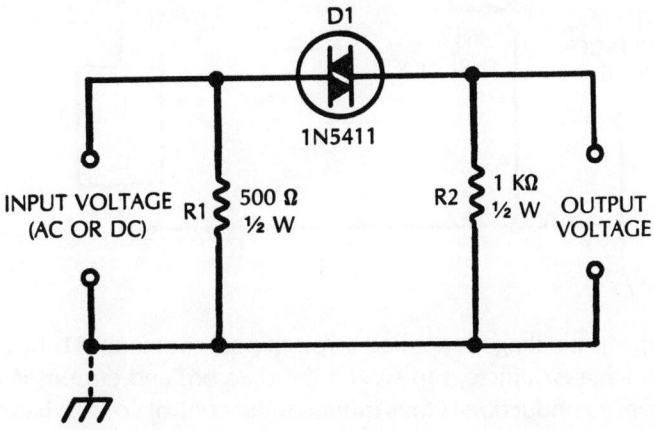

Fig. 11-3.

In operating the circuit, adjust the input voltage upward slowly from zero while monitoring the output. Up to approximately 30 volts, there will be little or no output voltage, owing to the very low leakage current of the diac. At approximately 35 volts, however, the diac will suddenly break down and an output voltage will appear across resistor R2. Reduce the input voltage, and note that the output voltage also decreases, eventually reaching zero when the input voltage is zero. At zero, the diac is "extinguished," and is in a condition to be triggered again by the 35-volt amplitude.

No. 107: Static dc Switch

The simple switch described in the preceding section may be triggered also by means of a change in voltage. Thus, a steady voltage of say 30 volts may be applied continuously to the 1N5411 diac without conduction taking place; but if an additional voltage of say 5 volts is added in series, the breakdown voltage of 35-volts is reached and the diac "fires." Removal of the 5-volt "signal" then will have no effect on the conduction which will continue until the 30-volt supply voltage is reduced to zero. This behavior is somewhat similar to that of a thyratron tube, but in this instance is obtained with the simple 2-terminal diac.

Figure 11-4 shows a switching circuit employing the principle of incremental-voltage switching described above. In this arrangement, a 30-volt bias is applied to the 1N5411 diac (D1) by the supply (while a battery is shown in Fig. 11-4 for simplicity, the 30 volts can be supplied by any other source of steady dc). With this voltage, the diac does not conduct, and no current flows through the external load.

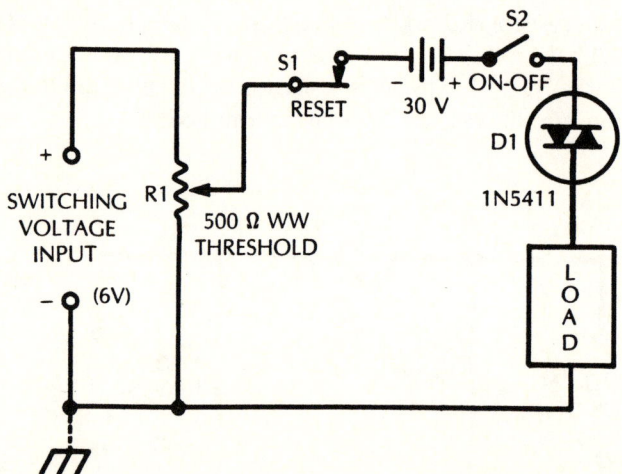

Fig. 11-4.

When the input voltage is applied through potentiometer R1, however, the increase in voltage is sufficient to switch the diac on, and current is delivered to the load. Once conduction is thus initiated, the control voltage has no further effect. Operating the reset switch, S1, temporarily interrupts the 30-volt supply and restores control to the switching voltage.

Most 1N5411s will be able to idle at 30 volts without self-firing. However, an individual unit may require a lower bias voltage. Similarly, most units will fire with the 6-volt increment shown in Fig. 11-4, but more sensitive units will operate with a smaller switching voltage which may easily be selected by means of the 500-ohm wirewound potentiometer, R1.

The maximum load resistance recommended for use in this circuit is 1000 ohms. The load current at this resistance is approximately 14.3 milliamperes.

This circuit will find use wherever a simple electrical latching action is desired without the complexity of 3-element thyristors and where the current demand is not severe. Individual diacs show surprising sensitivity in this simple arrangement.

No. 108: Electrically Latched Relay

Figure 11-5 shows the circuit of a dc relay that will remain closed (latched) once it has been actuated by a control signal. It has the dependability of a mechanically latched relay. This circuit employs the principle described in the preceding section; that is, the IRD54-C diac here is biased at 30 volts, a potential too low for conduction. But when a 6-volt increment is applied to the diac, the latter passes current which picks up the relay (the diac then continues to conduct, holding in the relay, although the 6-volt control voltage ceases).

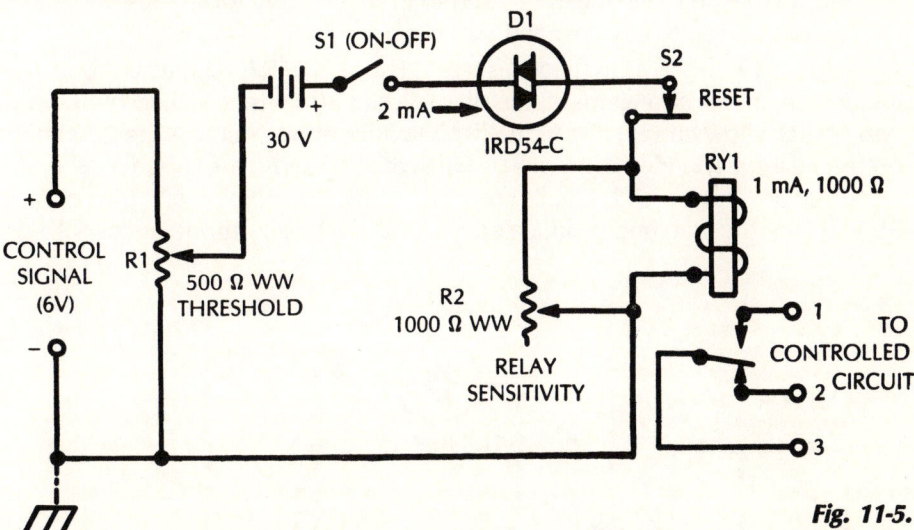

Fig. 11-5.

In this circuit, most IRD54-Cs will be able to idle at 30 volts without self-firing; however, an individual unit may require a lower voltage. Similarly, most units will fire readily with the 6-volt increment (control signal) shown in Fig. 11-5, but more sensitive units will operate with a lower control voltage which may easily be selected by means of the 500-ohm wirewound potentiometer, R1.

In its on state, the diac passes a current of 2 mA. The relay (RY1) is a 1 mA, 1000-ohm unit, so the extra 1 milliampere is diverted through resistor R2. Since the latter is a wirewound rheostat, it can be adjusted—along with potentiometer R1—for desired sensitivity of response.

While a 6-volt potential is required in series with the 30-volt bias to switch on the diac, a higher-voltage control signal may be employed, and R1 adjusted for the 6-volt value. The 30-volt supply is shown here as a battery, for simplicity, but can be any well-filtered dc source.

When R1 and R2 are properly set, the relay picks up readily when the control signal is applied to the input terminals of the circuit. Then, the relay continues to hold even when the control signal is removed, and holds until the reset switch, S2, is depressed momentarily. The relay is a 1 mA, 1000-ohm unit (Sigma 5F or equivalent) having ¼-ampere contacts. Use terminals 1 and 3 when the external circuit must be closed by relay closure; use 2 and 3 when the external circuit must be opened. If the ¼-ampere contact rating is too low for the power to be handled in the controlled circuit, a higher-wattage auxiliary relay may be operated from this circuit.

No. 109: Latching Sensor Circuit

Some systems, such as burglar alarms and process controllers, require an actuating signal that remains on after it has been initiated and ceases only when operation is reset from a central control point. Once available, this signal can be used to drive circuitry for alarms, recorders, shutoff values, safety devices, and so on.

Figure 11-6 shows a circuit of this type. Here, a HEP R2002 diac is the switching device. In this arrangement, the diac idles at 30 volts, supplied by B2, and passes such a low current that virtually no voltage appears across the 1000-ohm output resistor, R2. However, closure of switch S1, which may be a "sensor" on a door or window, adds 6 volts (from B1), to the 30-volt bias, and the resulting 35 volts fire the diac and produce approximately 1-volt output across R2.

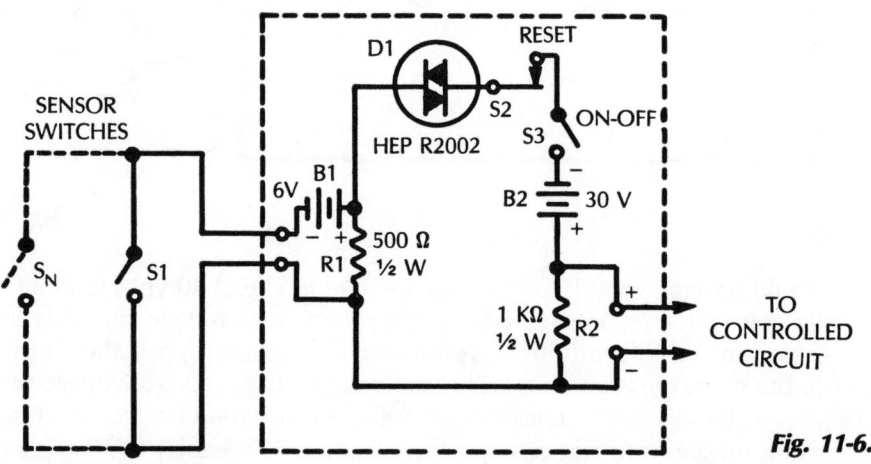

Fig. 11-6.

The diac then remains on in spite of any subsequent opening of S1 (as by closing of the door or window) until the reset switch (S2, at a secret location) is temporarily opened. Any number of "sensor" switches, such as S_n, may be operated in parallel, but this system requires that switches close when actuated, whereas the common burglar alarm system requires that they open.

While batteries are shown for simplicity, both the B1 and B2 voltages may be obtained from any other source of well-filtered dc. The 30 volts may be obtained with any convenient combination of batteries (one selection is two 15-volt units: Eveready 417, or equivalent).

For output voltage of the opposite polarity, simply reverse the connections to B1 and B2; the diac, being bilateral, requires no reversal.

No. 110: Dc Overload Circuit Breaker

Figure 11-7 shows a circuit that automatically disconnects a load device from its power supply when the dc operating voltage exceeds a predetermined value. The device then remains disconnected until the voltage is reduced and the circuit is reset.

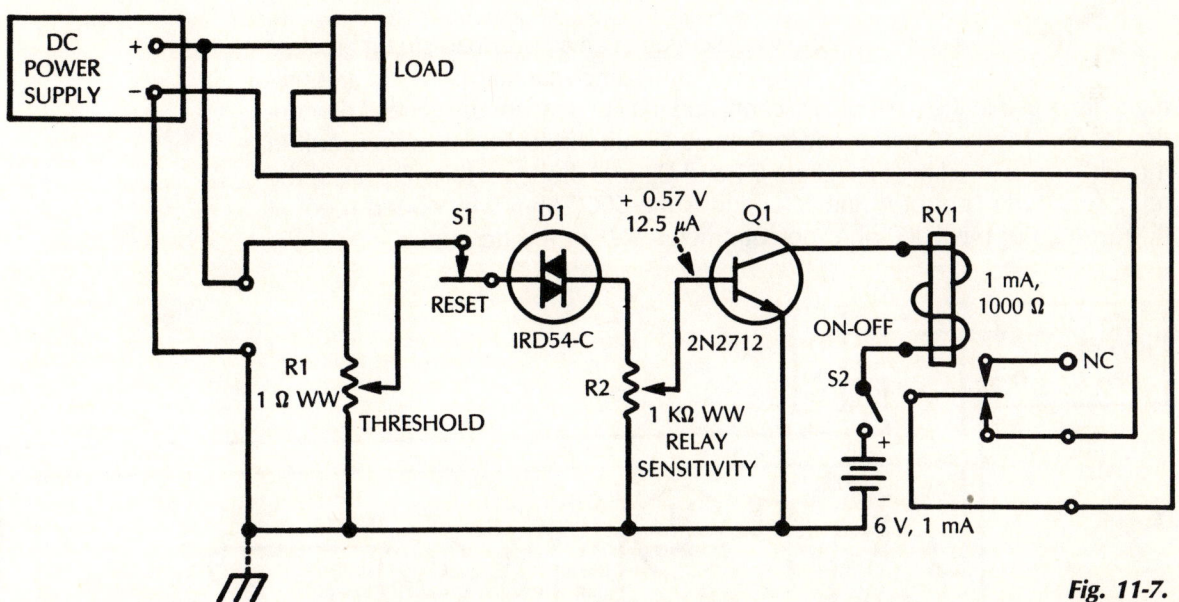

Fig. 11-7.

In this arrangement, employing an IRD54-C diac and 2N2712 silicon bipolar transistor, the diac (D1) is normally in its off state, and the static current of the transistor (Q1) is too tiny to actuate the sensitive relay (RY1). When the power supply voltage exceeds a critical value determined by the setting of potentiometer R1, the diac switches on and its dc output is applied to the transistor through potentiometer R2. The resulting increase in collector current actuates the relay which then opens the lead between the power supply and load. The diac then remains on and the relay actuated until the voltage returns to normal and the reset switch, S1, temporarily is opened.

To adjust the circuit initially, adjust potentiometers R1 and R2 so that the relay just operates when the power-supply voltage reaches the selected critical value. The relay then should remain actuated until the voltage falls back to its normal level and the reset switch is temporarily opened. When the circuit is operating correctly, the "firing" voltage at the diac input should be approximately 35 volts (individual diacs may switch on at a lower voltage, but this can be accommodated by adjustment of potentiometer R2), and the dc voltage at the base of the transistor should be approximately 0.57 volt (at approximately 12.5 μA). Individual transistors may require higher or lower voltage here, but this can be accommodated by means of potentiometer R2.

The relay is a 1 mA, 1000-ohm unit (Sigma 5F, or equivalent). As shown in the diagram, only two of the relay contacts are used, giving the relay a normally closed status. The relay contacts are rated ¼-watt; for higher power service, relay RY1 can actuate an auxiliary heavier duty unit.

No. 111: Ac Overload Circuit Breaker

Figure 11-8 shows the circuit of an ac overload circuit breaker. This circuit performs in the same manner as the dc arrangement described in the preceding section, and may be read for the general description of operation and performance. The ac circuit differs from the dc circuit in the addition of blocking capacitors C1 and C2 and diode rectifier D2; also, the relay sensitivity control in the ac circuit is the 5000-ohm wirewound rheostat (R3) instead of the 1000-ohm potentiometer (R2) in the dc circuit.

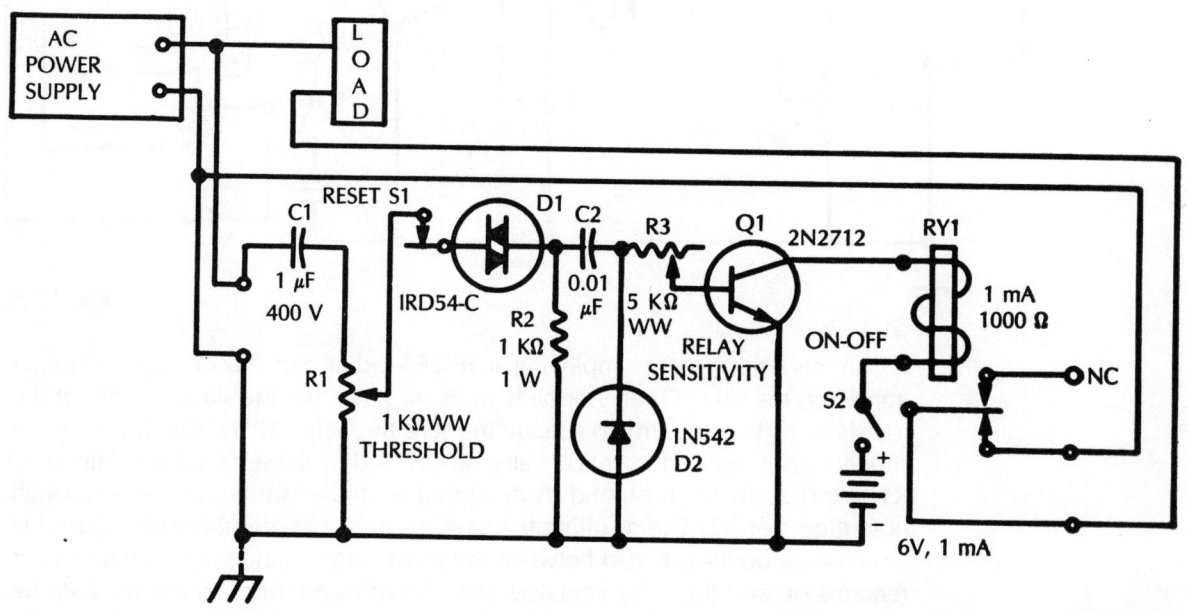

Fig. 11-8.

No. 112: Phase-Controlled Trigger Circuit

As mentioned earlier, the principal use of the diac at present is to supply a trigger voltage to a triac in various control circuits. The diac circuit for this purpose is phase controlled and by itself can find use in other applications than triac control, where a phase-adjustable pulse output is needed. Figure 11-9 shows the classic diac trigger circuit.

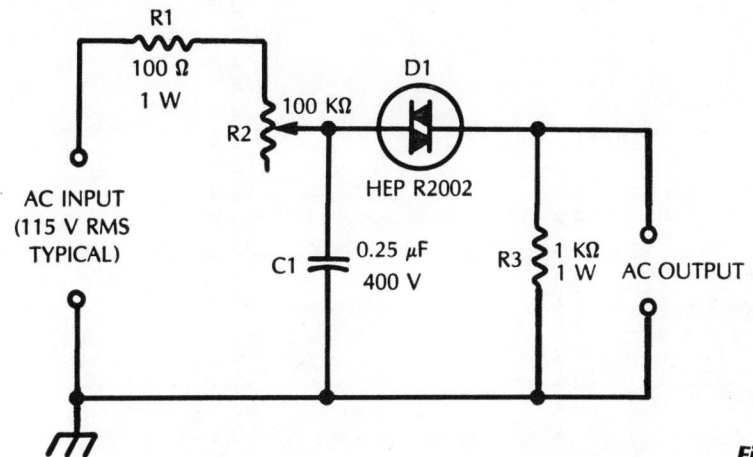

Fig. 11-9.

This arrangement essentially controls the angle of firing of the diac, and this is accomplished by adjustment of the phase-control network (R1-R2 and C1). The resistance and capacitance values given here are representative. For a given frequency (usually the power-line frequency), R2 is adjusted so that the breakover voltage of the diac is reached at the time instant corresponding to the desired point in the ac half-cycle at which the user desires that the diac switch on and deliver the pulse. The diac then will repeat this action during each half-cycle, positive and negative.

Ultimately, the phase is determined not only by R1-R2 and C1, but also to some extent by the impedance of the ac source and the impedance of the circuit which the diac arrangement triggers. For most purposes, however, it will be helpful to examine the phase of the diac-circuit resistance and capacitance to determine performance of the circuit.

Table 11-1, for example, shows the phase angles corresponding to various settings of the resistance against the 0.25 µF capacitance in Fig. 11-9. The data are given for 60 Hz, the frequency at which the diac trigger circuit is so often used. Note from these data that as the resistance is lowered, the trigger pulse appears earlier and earlier in the supply-voltage cycle, thereby enabling the diac to "fire" earlier in the cycle and to conduct longer. Because the RC circuit consists of series resistance and shunt capacitance, the phase is, of course, lagging—which means that the trigger pulse follows the supply-voltage cycle in time sequence.

f = 60 Hz RESISTANCE (R, OHMS)	C = 0.25 µF PHASE ANGLE (DEGREES LAG)
100	0.5
200	1.1
300	1.6
400	2.2
500	2.7
600	3.3
700	3.8
800	4.3
900	4.9
1000	5.4
2 k	10.7
3 k	15.8
4 k	20.7
5 k	25.2
6 k	29.5
7 k	33.4
8 k	37.1
9 k	40.3
10 k	43.3
20 k	62.1
30 k	70.5
40 k	75.2
50 k	78.1
60 k	80.0
70 k	81.4
80 k	82.5
90 k	83.3
100 k	84.0

Table 11-1. Phase Angles.

12
Triac

The triac is a 3-electrode device of the thyristor family, which can switch either ac or dc. Unlike the diac (Chapter 11), it has a separate control (gate) electrode which allows selection of the voltage level at which the triac begins to conduct. Like other thyristors, the triac behaves somewhat in the manner of the thyratron tube.

The triac finds application principally in control, switching, and trigger circuits. It is used singly or in conjunction with diacs, transistors, or silicon controlled rectifiers. The ratings of triacs cover a wide range, typical values being 100 volts to 600 volts and 0.5 amps to 40A.

THEORY

Figure 12-1A shows the basic internal structure of the triac, and Fig. 12-1B the circuit symbol. In this device, main terminals 1 and 2 are the output and common terminals, and the gate is the input or control terminal.

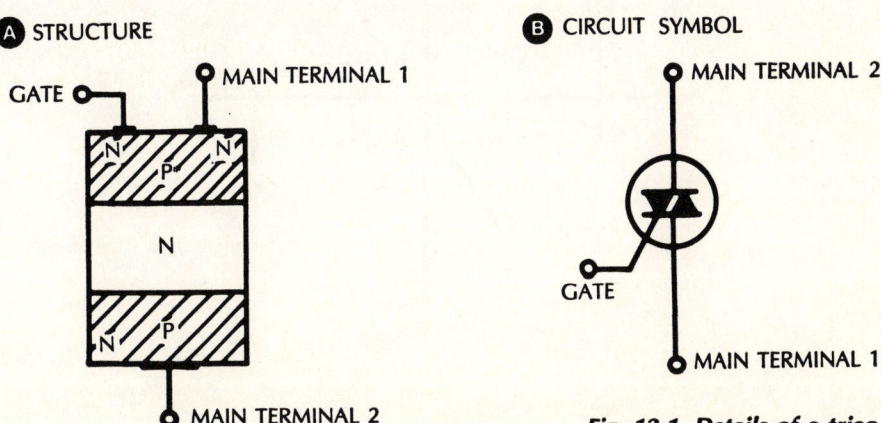

Fig. 12-1. Details of a triac.

Because of the series p and n layers in the triac pellet, this device—like the diac—can never pass current from terminal 1 to terminal 2 in the forward direction, regardless of the polarity of the applied voltage, but always behaves as a reverse-biased diode.

Figure 12-2 depicts performance of the triac. When a voltage is applied to this device, as from the main power source as shown in Fig. 10-2A, a very small leakage current flows (as from 0 to A, or 0 to D in Fig. 12-2B); this is the off state of the triac. As the voltage is increased, a critical value eventually is reached ($+V_{BO}$ in the positive direction, or $-V_{BO}$ in the negative direction) at which avalanche breakdown occurs and a heavy current suddenly flows (B to C and beyond, or E to F and beyond); this is the on state of the triac. The breakover voltage ($+V_{BO}$ or $-V_{BO}$) is determined by the amplitude of a positive or negative current pulse supplied to the gate electrode; the higher this amplitude, the lower the breakover voltage.

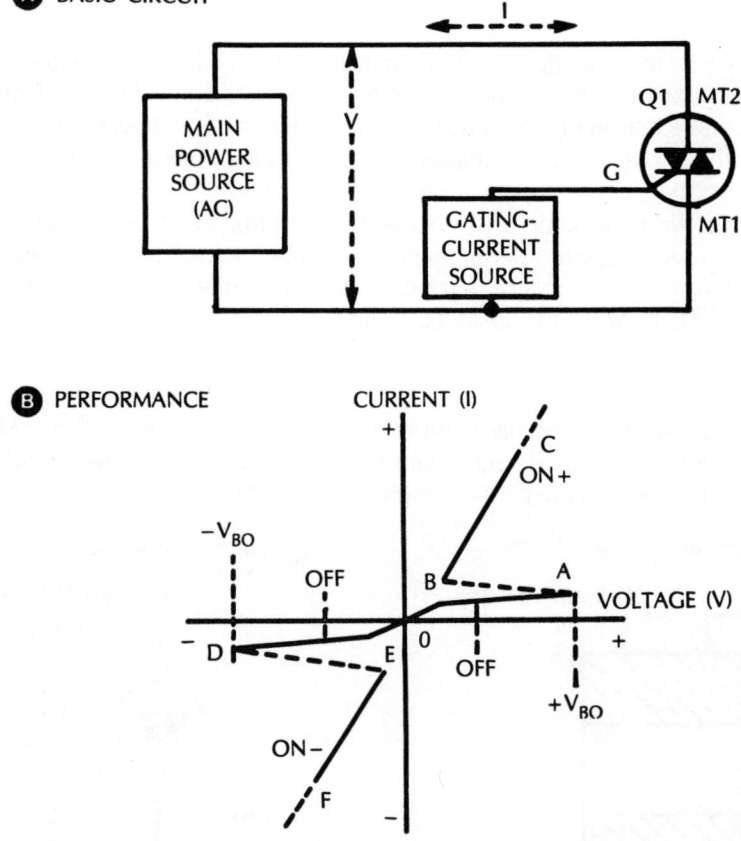

Fig. 12-2. Triac action.

Figure 12-2B shows that the triac snaps to its on state, positive or negative. As in a thyratron tube, once dc conduction has been initiated in the triac, the gate electrode under ordinary conditions exerts no further control until the voltage from terminal 1 to terminal 2 is interrupted or otherwise reduced to zero.

Unlike the diac, the triac has exclusive terminals which must not be interchanged. Thus, in the circuit diagrams in this chapter, Main Terminal 1 is labeled MT1, Main Terminal 2 is labeled MT2, and the gate is labeled G. The gate electrode is at the Main Terminal-1 end of the triac structure (see Fig. 12-1A) and is so indicated in the circuit symbol (see Fig. 12-1B and the circuit diagrams in this chapter).

Some Triacs may be worked harder than usual if a heat sink is provided for them. An immediate example is the triac motor control: In some instances, a ¼-horsepower motor is the largest machine that can be accommodated by a certain control circuit; but if the triac is provided with a suitable heat sink, a ½-horsepower motor can be safely controlled.

For light dimmers, RCA advises that with small units built into the socket of the controlled lamp and handling up to 2 amps rms, sufficient heat sinking can be obtained from the lead wires to which the triac is attached. But up to 6 amps in a wall-mounted unit, the triac must be fastened to the metal wall box and face plate for sinking, and for still higher current, the face plate must be of a type that can be mounted clear of the wall for air flow underneath.

No. 113: Simple Triac Switch

In the simplest application of triac theory, an ac-powered triac may be turned off by removal of the gate current. In this way, a large ac current (amperes) can be switched on and off through a load by means of a very small gate current (milliamperes). The ratio of controlled power (a characteristic somewhat similar to power amplification) can have a value of several thousand to one. The load can be any device—such as a motor, lamp, or heater—operating within the current rating of the triac, that ordinarily would be operated by means of a switch, heavy-duty relay, high-current thermostat, or similar control.

Figure 12-3A shows the circuit of a simple switch employing a 2N5754 triac. The latter is rated at 2.5 amps, and a heavier-duty triac may be used in its place if higher-current operation is desired. The adjustable trigger current is supplied by the resistance combination R1-R2 connected back to the supply voltage. Rheostat R1 is a 200,000-ohm, 1-watt, linear-taper unit (Mallory Midgetrol or equivalent). Fixed resistor R2 protects the rheostat from direct connection to the high voltage. Normally open switch S2 (SPST) is the triggering device. Instead of a simple switch, however, the light-duty contacts of a sensitive relay, a photoconductive cell, or a temperature sensor might be used. Closing the contacts or reducing the resistance of the sensor passes approximately 10 mA into the triac gate section and switches the triac on to pass up to 2.5 amps through the load.

Figure 12-3B depicts operation of the circuit. The plot shows the angle of start and the duration of main current flow. Because of the simple resistive control circuit, the gate current is in phase with the triac voltage. When the gate current is low at the beginning of the ac cycle, however, the triac acts as a very high resistance and virtually no current flows through the load.

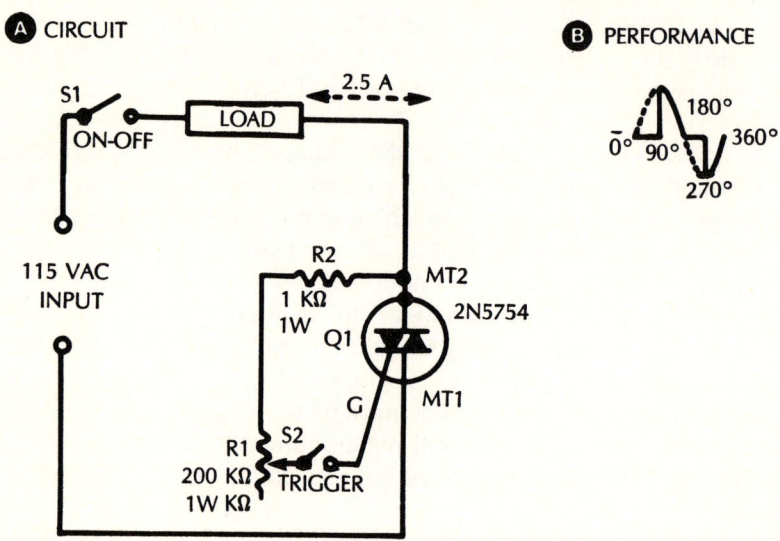

Fig. 12-3. Simple triac switch.

When R1 is appropriately set for a given triac and ac voltage, the gate current reaches the trigger value for the 2N5754 when the supply voltage is at the maximum point in the cycle, and the triac is abruptly switched on. Conduction then continues until the end of the positive half-cycle, whereupon Q1 switches off as the supply-voltage cycle passes through zero, and remains off until the maximum-voltage point in the negative half-cycle is reached. At that point, Q1 again switches on and remains on until the zero point at the end of the cycle. Thus, the trigger voltage and current determine the instant at which the triac is switched on and the interval of current flow through the load, in this instance the flow being from 90° to 180° and from 270° to 360°.

If rheostat R1 is adjusted to a lower resistance, the trigger-voltage value will be reached earlier in the cycle and the load current will flow for a greater part of each half-cycle. This is essentially an off-on circuit and might invite the question of why it should be used in lieu of a simple switch, such as S1. The answer is that a switch such as S1 must handle the full 2.5 amps, whereas S2 (in whatever form) need handle only a few milliamperes. Any pair of light-duty contacts—or equivalent device—can serve as S2.

No. 114: General-Purpose Controller

Figure 12-4 shows the circuit of an ac controller which, though very small by comparison, may be used in the manner of a Variac to adjust the ac input to any device connected to the load socket that is capable of control by a triac. The 40429 triac shown here provides a maximum output current of 6 amps and may be operated up to 200 volts. The range of control, from maximum output to very nearly complete cutoff, is provided by a simple, volume-control-type rheostat (R1).

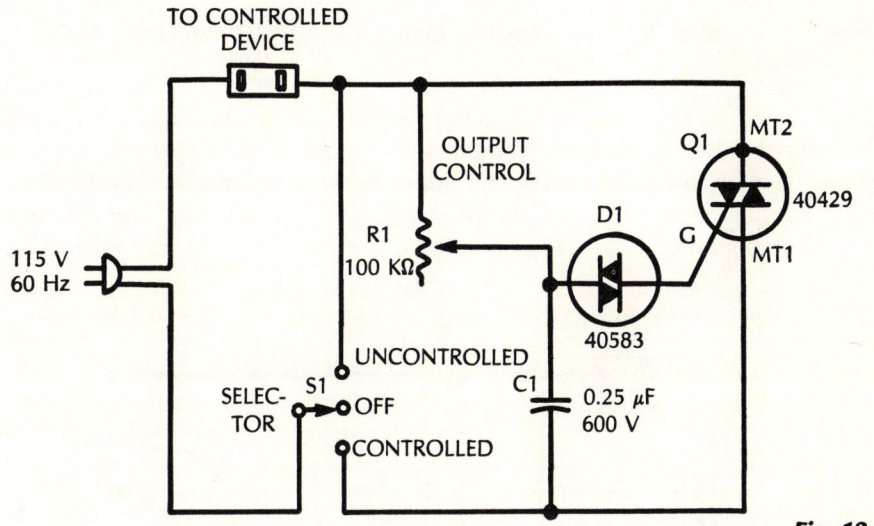

Fig. 12-4.

The triac is triggered by means of a 40583 diac (D1) operated from a "tuned" phase-shift network, R1C1. (See Chapter 9 for a general discussion of the diac trigger circuit.) The point in the ac supply-voltage cycle at which the diac triggers the triac is determined by the setting of R1 which, in turn, adjusts the phase of the trigger with respect to that of the supply voltage.

In this way, the rheostat setting determines the angle of flow and consequently the output current (Once the triac has been triggered, output current continues to flow until the end of that half-cycle, and is retriggered at the corresponding point in the succeeding half-cycle. See Fig. 12-3B). The lower the resistance setting of R1, the higher the output current of the circuit.

Because no practical capacitor (C1) can give the full 90° phase shift attributed to capacitance, the phase shift of the R1C1 network is never sufficient to cut the triac completely off; nevertheless, the control range is wide and very useful. For complete turn-on and turn-off, switch S1 acts as a mode selector. In the uncontrolled position of this switch, the full line voltage is applied to the load without benefit of the controller circuit which then stands idle. With S1 in the controlled position, the load is supplied with adjustable current delivered by the triac. With S1 in its off position, the load is deenergized.

For heavier-duty service than that afforded by the 6-amp triac shown in Fig. 12-4, higher-powered triacs are available. Representative examples are Types 40668 (8-amps, 200-volts), 2N5567 (10-amps, 200-volts), 2N5568 (10-amps, 400-volts), 2N5571 (15-amps, 200-volts), 40707 (30-amps, 200-volts), and 2N5444 (40-amps, 200-volts). Each may be triggered with the same R1C1D1 circuit shown in Fig. 12-4.

Some of the familiar uses of the circuit include control of lamps (except fluorescent), motors (universal and induction), electric heater or oven elements, soldering irons, and ac-powered electronic apparatus.

No. 115: Controller Using Combination Thyristor

The controller circuit shown in Fig. 12-5 operates on the same principle as that of the general-purpose controller shown in Fig. 12-4, and the description of operation and applications in the preceding section applies as well to Fig. 12-5. The difference is that the latter circuit employs a 40431 combination thyristor (Q1) which contains both the diac and triac units. This combination unit somewhat simplifies assembly, wiring, and replacement.

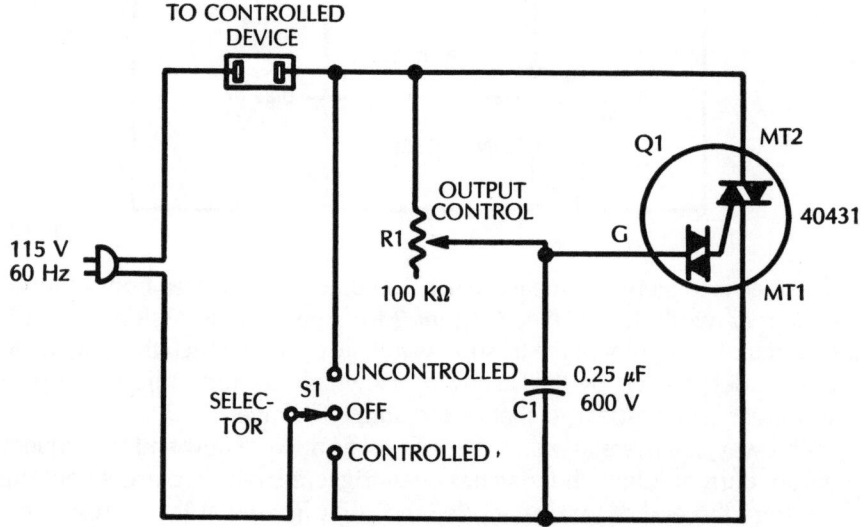

Fig. 12-5.

The 40431 thyristor is rated at 6-amps, 200-volts. A similar unit—Type 40432—is rated at 6 A, 400 V. Type HEP R1725 is equivalent to 40431, and HEP S0015 is equivalent to 40432. All may be operated with either resistive or inductive loads.

No. 116: Dc-Controlled Solid-State ac Relay

In electrical and electronic installations, it is often necessary to switch high ac voltage and current with a low-current dc control signal. At least two electromechanical relays operating in cascade usually are required for this purpose, since the contacts of a sensitive (low-voltage, low-current) dc relay cannot handle the high ac current.

For this application, an all solid-state relay circuit—with no moving parts whatever—may be achieved with a suitable triac, since a triac may be switched with a dc gate trigger of either polarity and requires only a few milliamperes to switch several amperes.

Figure 12-6 shows such a circuit. The 40773 triac (Q1) shown here is rated at 2.5-amps, 200-volts. A similar unit—Type 40774—is rated at 2.5-amps, 400-volts. In this arrangement, the triac is triggered on by a dc gate voltage (1.5 to 2 volts at 20 milliamperes) applied to the dc control-signal input terminals.

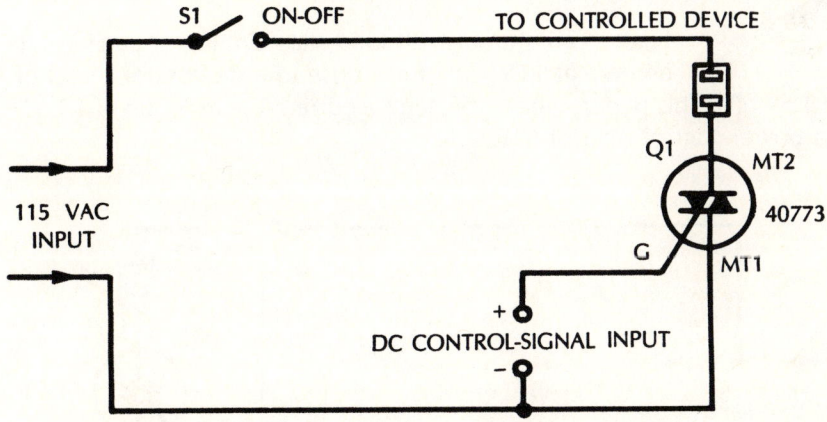

Use Only with 1:1 ratio, 115 V Isolating Transformer

Fig. 12-6.

The trigger-voltage and current values may differ somewhat from these figures with individual triacs.

Once the triac has switched on, it will continue to conduct for the remainder of the ac half-cycle, extinguishing as the supply-voltage cycle crosses the zero line, and retriggering at the corresponding point in the succeeding half-cycle. The ac will continue to flow through the load as long as the dc control signal is present at the gate, and will cease when the dc is interrupted. While the dc control-signal input terminals are labeled + and − in Fig. 12-6, the polarity shown is not mandatory; the triac triggers on either a positive or a negative gate voltage.

This relay circuit allows 2.5 amps at 115 Vac to be switched with approximately 20 mA at 1.5-2 Vdc. This represents a current-control ratio of 9583:1 when the dc control voltage is 1.5. For increased sensitivity, a transistor or IC direct-current amplifier may be operated between a millivolt/microampere type of dc signal source and the dc control-signal input terminals (see, for example, Fig. 12-7).

Because this type of relay circuit is often operated from an external dc signal source which must be returned to main terminal 1 of the triac as shown in Fig. 12-6, this source will automatically be connected dangerously to one side of the ac power line. Therefore, to avoid electric shock or damage to the equipment, it is advisable to insert a 1:1 isolating transformer between the power line and the circuit in Fig. 12-6.

When this solid-state relay must switch higher current than the 2.5-amp rating of the 40773 triac, employ a heavier-duty triac at Q1. Higher-current triacs include Types 40733 (4.2-amps, 200-volts), 40429 (6-amps, 200-volts), 40668 (8-amps, 200-volts), 2N5567 (10-amps, 200-volts), 40662 (30-amps, 200-volts), and 40688 (40-amps, 200-volts).

No. 117: Sensitive Dc-Controlled Triac Switch

Figure 12-7 shows an all solid-state circuit that can switch 0.5 ampere at 115-Vac in response to a dc control signal of 47 µA at 1.5-volts. This performance represents a current-control ratio of 10,638:1, and a power-control ratio of 815,603:1.

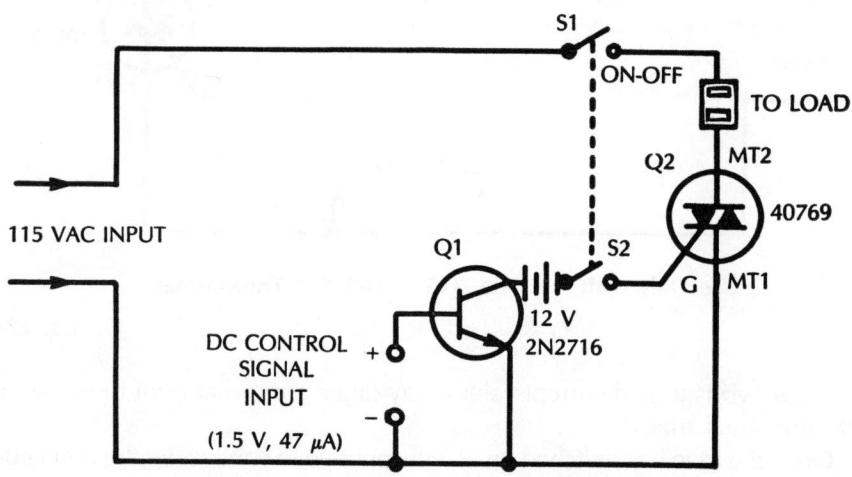

Use Only With 1:1 Ratio, 115 V Isolating Transformer

Fig. 12-7.

In this circuit, the dc control signal is amplified by a single-stage dc amplifier based upon the 2N2716 silicon bipolar transistor (Q1) to an output current of 10 mA which is applied as a gate current to trigger the 40769 triac (Q2) into conduction. The triac conducts ac through the load as long as the dc is present at the dc control-signal input terminals, and ceases when the dc is interrupted.

The 2N2716 provides very low-drift operation, insuring against self-triggering of the circuit.

With individual transistors and triacs, the dc signal requirement may differ somewhat from the 1.5-volts and 47 µA indicated in Fig. 12-7. The dc control signal can be obtained from a rectifying diode, self-generating photocell, photoconductive cell, thermocouple, light-duty contacts making and breaking a 47 µA input current, and so on.

The floating 12-volt transistor dc supply is shown here as a battery, and a small battery will be preferred in many installations; however, a well-filtered, power-line-operated supply may be used instead. This type of relay circuit is often operated from an external dc signal source which must be returned to the emitter of the transistor and main terminal 1 of the triac, as shown in Fig. 12-7. The source thus will be automatically connected dangerously to one side of the ac power line. Therefore, to avoid electric shock or damage to the equipment, it is advisable to insert a 1:1 isolating transformer between the power line and the circuit in Fig. 12-7.

When this solid-state relay must switch higher current than the 0.5-amp rating of the 40769 triac, use a heavier-duty triac at Q2. Higher-current triacs include types 40767 (1.6-amps, 100-volts), 40509 (2.2-amps, 200-volts), 40691 (2.5-amps, 200-volts), 40733 (4.2-amps, 200-volts), 40429 (6-amps, 200-volts), 40668 (8-amps, 200-volts), 2N5567 (10-amps, 200-volts), 2N5571 (15-amps, 200-volts), 40662 (30-amps, 200-volts), and 40688 (40-amps, 200-volts).

For use with negative control voltage, substitute a pnp transistor for the 2N2716, and reverse both B1 and the dc control-signal input terminals.

MOTOR CONTROLS

Diac/triac circuits are very popular for controlling the speed of electric motors. The circuits given in Figs. 12-4 and 12-5 are ideal for this purpose. As the resistance setting of rheostat R1 in each of these circuits is reduced, the angle of conduction of the triac increases and so does the motor speed. These circuits will not operate satisfactorily with motors of the capacitor-start, synchronous, repulsion-induction, and shaded-pole types, but are entirely useful with universal and induction motors. Both circuits will handle motors rated up to ¾ horsepower.

Nos. 118 and 119: Two Improved Light Dimmers

One familiar application of the diac/triac ac controller is the continuously variable adjustment of the brightness of incandescent lamps. This function, called light dimming, can be performed by any of the common circuits, such as Figs. 12-4 and 12-5. However, each of these simpler circuits exhibits a hysteresis effect which can be troublesome in some installations; that is, the controlled lamp switches on at a particular setting of the control rheostat when the latter is being turned up, but switches off at a somewhat different setting when the rheostat is being turned down.

This hysteresis can be reduced by employing two RC timing circuits in cascade, in place of the single circuit (R1-C1) in Figs. 12-4 and 12-5, an arrangement which provides two time constants. This improvement is shown in the light-dimmer circuits in Figs. 12-8 and 12-9. In each of these arrangements, adapted from a basic RCA circuit, the first timing leg is R2-C2, and the second leg is R3-C3.

During triggering, capacitor C3 discharges into the triac gate; but C2, being charged to a higher voltage than is C3 and having a longer discharge time, is able to replenish the C3 charge to some extent, and this reduces the hysteresis and extends the adjustment range of dimmer-control rheostat R2. The rheostat is a 200,000-ohm, 1-watt, linear-taper unit (Mallory Midgetrol or equivalent).

The two improved circuits are functionally identical. However, Fig. 12-8 employs a separate triac (type IRT82) and diac (type IRD54-C), whereas Fig. 12-9 employs a single thyristor (type 40431) which incorporates both diac and triac in the same package. Where needed, the latter arrangement somewhat simplifies construction, wiring, and replacement. The circuit given in Fig. 12-8 may be used in either 115- or 220-volt service (up to 6 amps). The circuit in Fig. 12-9 is intended solely for 115-volt service, but may be modified for 220-volt service by substituting a type 40432 unit for 40431.

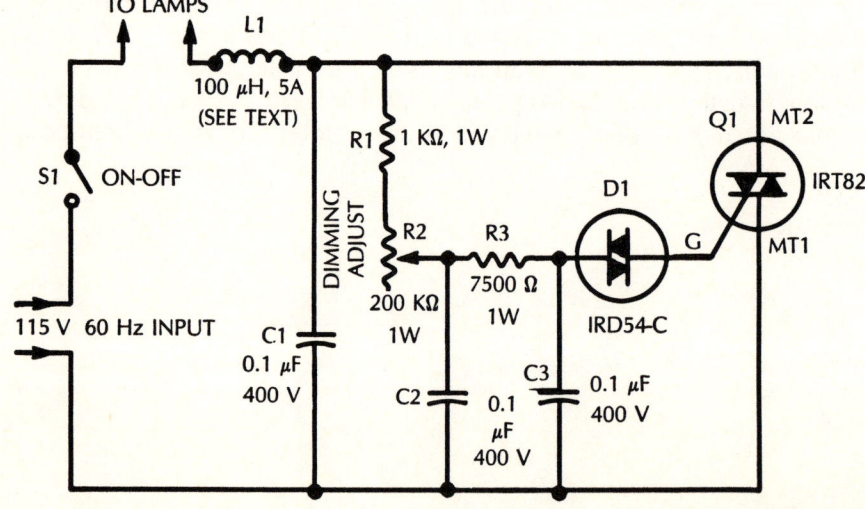

Fig. 12-8. Improved light dimmer.

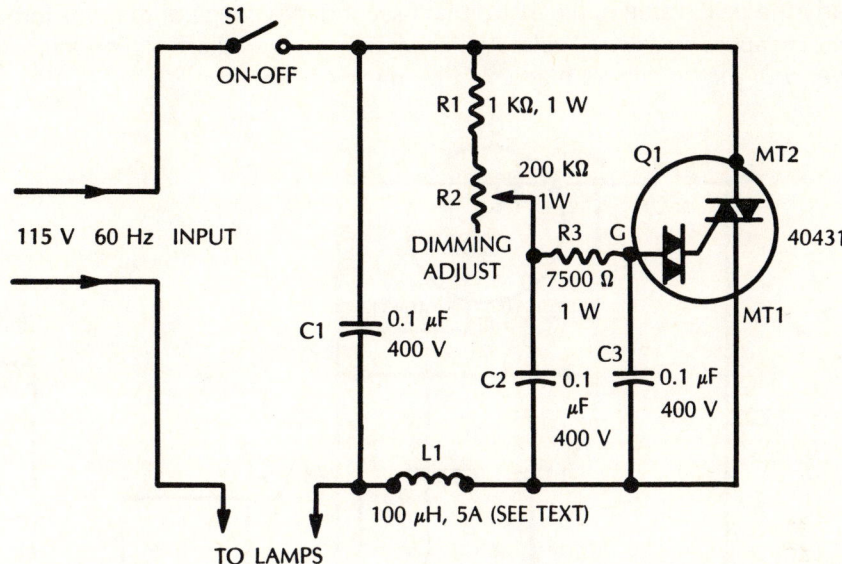

Fig. 12-9. Further improved light dimmer.

A further refinement in the improved light-dimmer circuits is the inclusion of a low-pass filter (L1-C1) to suppress radio/TV interference which might otherwise result from the rapid triggering of the triac. This filter requires a high-current (5-amp) 100 μH inductor of the smallest obtainable size (Dale IH5-100 or equivalent).

No. 120: Variable Dual, dc Power Supply

In Fig. 12-10, a continuously variable, dual-output, dc power supply is obtained by using a triggered triac ac controller to vary the primary current of the power transformer, T1. The circuit delivers two separate maximum dc voltages: +17-volts and −17-volts, each at a maximum of 80 mA. The controller employs a type 40431 combination thyristor, Q1 (diac and triac in the same package), and its circuit is similar to the one in Fig. 12-9. Rheostat R2 in the controller section of the power supply permits smooth adjustment of the dc output voltage, and is a 200,000-ohm, 1-watt, linear-taper unit (Mallory Midgetrol or equivalent).

In the power supply section, transformer T1 (Stancor P-6377 or equivalent) has a 24-volt center-tapped secondary. The rectifier, RECT1 (type HEP RO801), although built as a bridge rectifier, is not used as a bridge in this circuit. Instead, each two-diode half of the rectifier provides a full-wave, center-tap rectifier circuit in conjunction with the center-tapped secondary of the transformer, the right half delivering positive voltage, and the left half negative voltage. The 2200 μF, 35-volt electrolytic capacitors (C3 and C4) provide adequate filtering for most low-current service. At low output-current drain, these capacitors charge to very nearly the peak value of the 12-volt half-secondary voltage of the transformer, that is, to approximately 17 volts. At higher currents, the voltage is proportionately lower, but may easily be readjusted by means of rheostat R2.

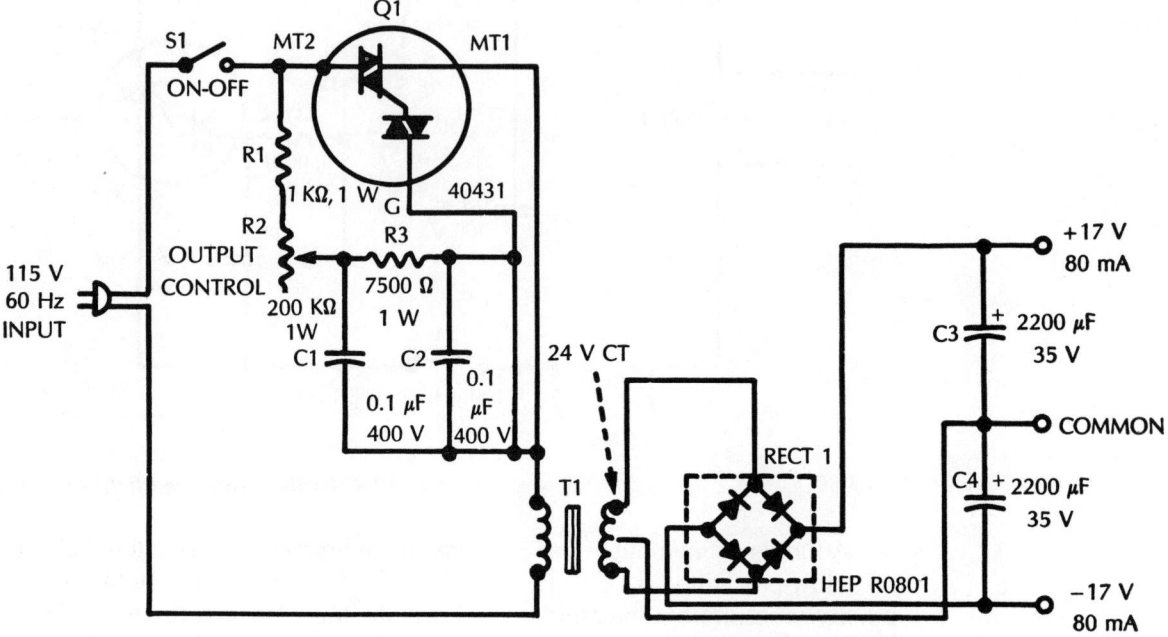

Fig. 12-10. Variable, dual-dc power supply.

A dual supply of this type is useful for generating adjustable plus and minus dc voltages for integrated circuits during circuit development and circuit or IC testing, and for similar applications requiring two identical dc voltages of opposite polarity. The same arrangement may be employed for other maximum output voltage and current values simply by appropriately changing the transformer, rectifier, and two electrolytic capacitors.

No. 121: Automatic Equipment Power Switch

Figure 12-11 shows how a triac may be used to switch off the power to a main unit (such as a master amplifier or main transmitter) automatically when the power to an auxiliary unit (such as a preamplifier or modulator) is manually switched off. This arrangement forestalls the accidental leaving on of a main unit, through forgetfulness, when the auxiliary unit is switched off. The method is due to T.N. Tyler (see *Popular Electronics,* January 1973, p. 52), and has been adapted here for readily available components.

In this arrangement, a HEP R1723 triac (Q1) is connected in series with the main-unit socket (P2) and the ac power line. The auxiliary unit is plugged into socket P1. As long as the auxiliary unit is operating, it draws its power through diodes D1 to D4, developing a voltage drop across these diodes; and this voltage triggers the triac on, allowing the main unit to receive power.

When later, the auxiliary unit is switched off, the diode voltage drop disappears, the triac switches off, and the main unit likewise is switched off even though its switch may absent-mindedly have been left on. If more than one auxiliary unit is used, separate sockets for them—such as P_N—may be connected in parallel with P1 and P2.

This "equipment minder" works well when the plugged-in units contain no reactive paths which can draw significant current through the diodes even when the units are manually switched off. Such a reactive path might be provided by a large suppressor capacitor connected across the power-line-input terminals inside the unit. The resulting capacitor current would maintain a voltage drop across the diodes and this would keep the main unit running in defeat of the whole idea. In some instances—but not all—such capacitors may be removed without degrading performance of the equipment.

The HEP R1723 triac shown in Fig. 12-11 is a 6-amp, 200-volt unit. A heavier-duty triac may be selected for higher power drain by the main unit.

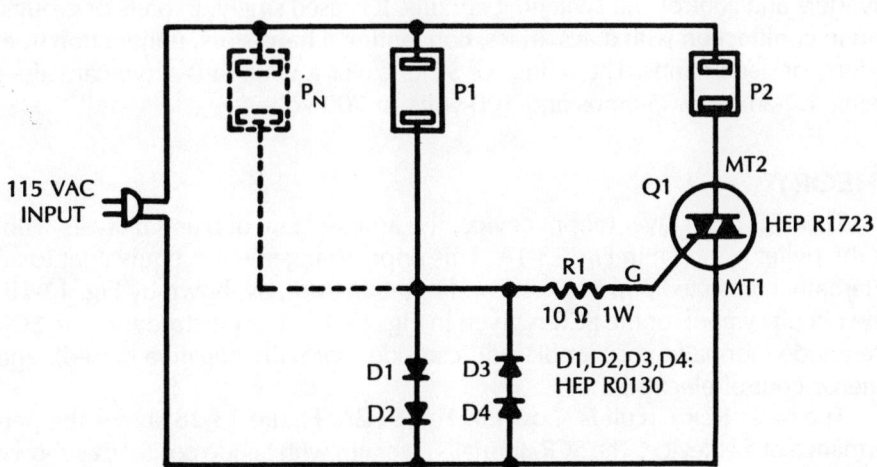

Fig. 12-11.

13
Silicon Controlled Rectifier SCR

The silicon controlled rectifier (SCR) is a 3-electrode device of the thyristor family, which can switch or control dc and rectify ac with controllable angle of conduction. Like the triac, it has a separate gate (control) electrode; and like other thyristors, the SCR behaves in a manner analogous to that of the thyratron tube—even more so than does the triac. Unlike the triac, the SCR can conduct in only one direction; its anode must be made positive and its cathode negative.

The SCR finds application principally in controlled rectification and in inverters and control and switching circuits. It is used singly, in pairs or groups, and in conjunction with diacs, triacs, conventional transistors, unijunction transistors, or neon lamps. The ratings of SCRs cover a wide range, typical values being 1.7-amps to 35-amps and 100-volts to 700-volts.

THEORY

The SCR is a 4-layer (pnpn) device; the arrangement of p and n layers within the pellet is shown in Fig. 13-1A. This pnpn arrangement is equivalent to an internally connected pnp transistor and npn transistor, as shown by Fig. 13-1B. The circuit symbol for the SCR is given in Fig. 13-1C. The electrodes of the SCR are anode (normally positive biased), cathode (normally negative biased), and gate or control electrode.

The basic SCR circuit is shown in Fig. 13-2A. Figure 13-2B shows the performance of the device. The SCR normally operates with anode positive, as shown here. If the anode is biased negative with respect to the cathode, only a small leakage current flows from D to E (as in a reverse-biased silicon diode).

Silicon Controlled Rectifier SCR

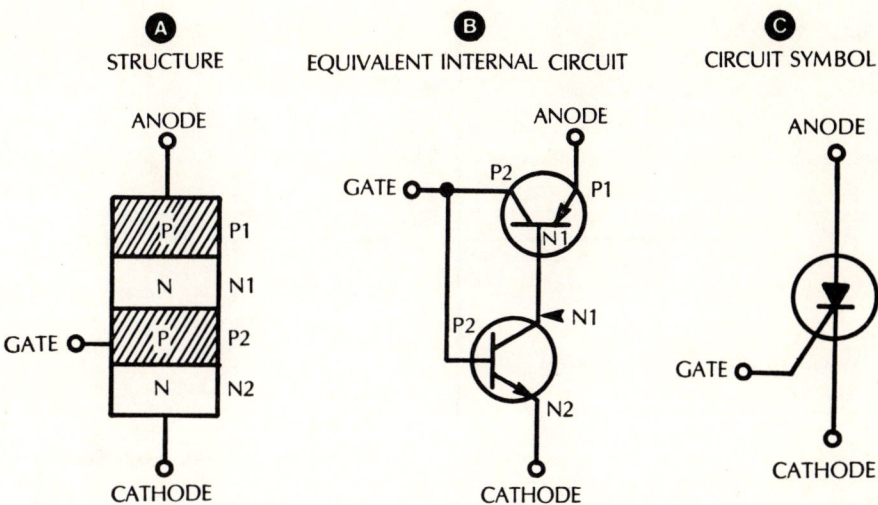

Fig. 13-1. Details of an SCR.

When the reverse breakdown voltage—V_B—is reached, however, a large, potentially destructive current flows, as from E to F and beyond. When the anode is properly positive, a very small leakage current flows (as from O to A); this is the off state of the SCR. When the anode voltage reaches the breakover voltage (V_{BO}) at A, the current, due to avalanche breakdown, increases sharply (B to C and beyond); this is the on state of the SCR. The current at point I_h is the holding current. The breakover voltage (V_{BO}) is determined by the value of the positive gating voltage applied to the gate electrode; the higher the gating voltage, the lower the breakover voltage, and vice versa. When the gate voltage is zero, the SCR blocks current in both directions (off state).

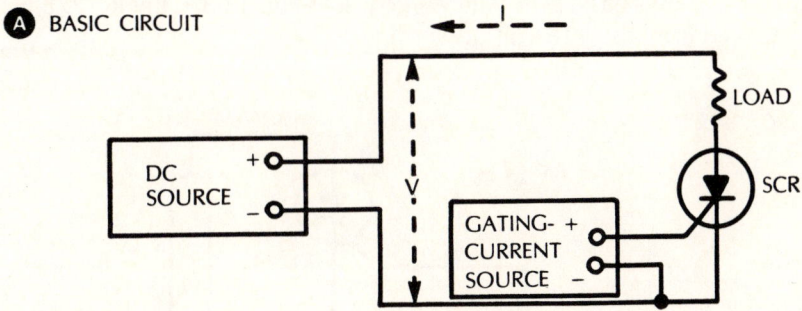

From the curve in Fig. 13-2B, it is seen that the SCR snaps to its on state.

As in a thyratron tube, once conduction has been triggered in the SCR, the gate electrode—under ordinary conditions—exerts no further control of anode current until the anode-to-cathode is interrupted or temporarily reduced to zero, whereupon control is restored to the gate.

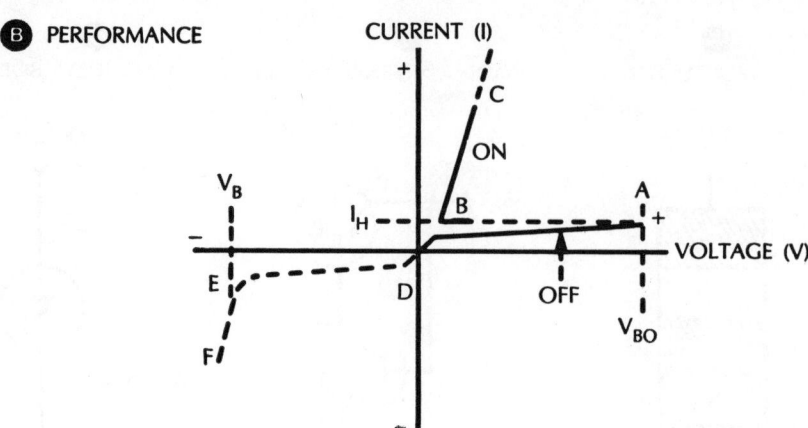

Fig. 13-2. SCR action.

Since the SCR, unlike the triac, is not a bidirectional device, it will cut off automatically and restore control to the gate at each reversal of the cycle when an ac voltage is applied to the anode (as in controlled rectification).

Nos. 122 and 123: Basic SCR Switches—Ac and Dc

With a silicon controlled rectifier, a large anode current may be switched by means of a small gate current. Thus, a 5 mA gate current will switch 2.5 amps in a type 40810 SCR, and 8 mA will switch 35 amps in type 2N3650. The gate current can be obtained from any of a number of different sources, and many types of load devices can be operated by the SCR. This is the simplest application of silicon controlled rectifiers.

Figure 13-3 shows basic SCR static switches. In Fig. 13-3A, the IR122B SCR (Q1) is operated from the 115-volt power line.

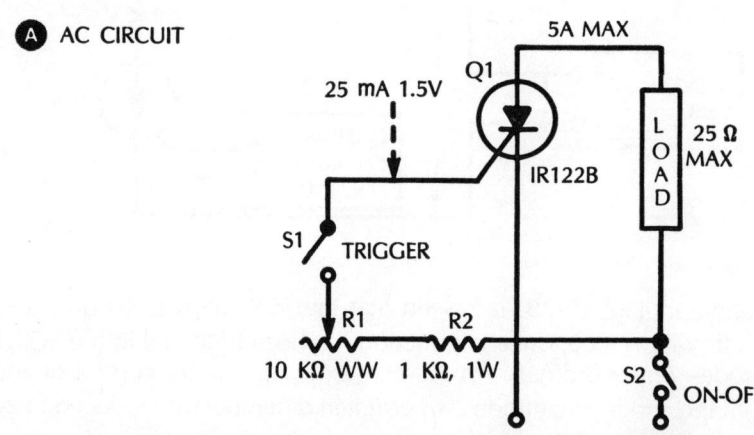

The ac gate current is obtained from the line through resistors R1 and R2, accordingly, is in phase with the anode voltage. Because the SCR is a rectifier, it conducts current only during the positive half-cycles of anode voltage, and it can conduct only when the gate current has reached a critical amplitude for that type of SCR.

In this circuit, if rheostat R1 is set so that the gate current reaches its critical trigger value at the instant that the positive half-cycle of anode voltage reaches its maximum value (90°), the SCR will fire at 90° and will remain in conduction—even if switch S1 is opened—until the end of the positive half-cycle (180°). At this point, the anode voltage being zero, the SCR switches off and remains off throughout the negative half-cycle (180° to 360°), since the SCR cannot conduct when its anode is negative.

During the succeeding positive half-cycle, however, the SCR is again triggered on at the 90° point (if the trigger voltage is still present at the gate) and conducts to the end of that half-cycle (180°). This performance is depicted by Fig. 13-3B, which shows that the anode current flows during the last half of each positive half-cycle. If R1 is set so that the gate current reaches its trigger value earlier in the ac half-cycle, then the SCR will switch on earlier and anode current will flow during a longer part of the positive half-cycle. In this ac circuit, once the SCR switches off, when the ac cycle passes through zero, control is restored to the gate.

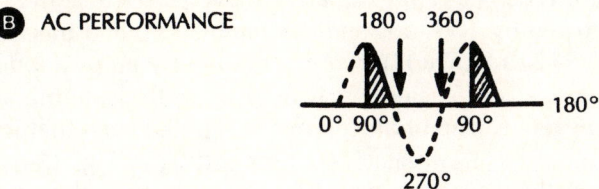

B AC PERFORMANCE

In Fig. 13-3C, the IR106Y1 SCR (Q1) is operated from a 35-volt dc source; otherwise, the circuit is of the same type as the one in Fig. 13-3A.

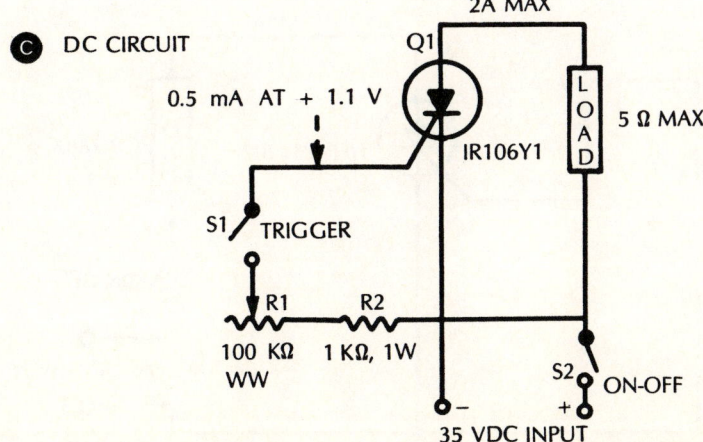

C DC CIRCUIT

Fig. 13-3. Basic SCR switches.

In the dc circuit, if rheostat R1 is set so that the gate current has the critical trigger value required by that type of SCR, Q switches on when S is closed. The SCR then conducts anode current through the load and will remain on—even if S1 is opened—until the dc anode voltage is interrupted, as by momentarily opening switch S2. Thus, S2 always must be opened to reset the circuit to respond to subsequent trigger signals.

The advantage of the SCR switch, ac or dc, is that it allows a low current and voltage to be used to switch a much higher current and voltage. Thus, in Fig. 13-3A, a gate current of 25 mA at 1.5-volts switches 5-amps at 11-volts; and in Fig. 13-3C, 0.5 mA at 1.1-volts switches 2-amps at 35-volts. Various SCRs afford other levels of sensitivity and other maximum anode voltages and currents. Because of the comparatively low gate current and voltage, the contacts of the control switch, S1, may be light. Also, S1 need not be a switch per se, but may be light-duty relay contacts or in some applications a thermostat, photocell, thermocouple, humidity sensor, thermistor, or other similar device that will deliver a low dc voltage.

No. 124: Light-Controlled SCR

Figure 13-4 shows how the dc output of a self-generating silicon photocell can supply the gate trigger current to an SCR. The type S6M-C photocell (PC1) delivers approximately 1.5 volts (at a maximum current of 8 mA) when actuated by 100 footcandles illumination, and this dc output will trigger the HEP R1220 SCR (Q1) to a maximum of 2-amps anode current through the load. Either ac or dc supply may be used; with dc, terminal A must be positive, and B negative. The photocell must be poled such that its positive dc output is applied to the SCR gate (see color coding in Fig. 13-4).

To set up the circuit initially, illuminate the photocell and adjust rheostat R1 to the point at which the SCR switches on and passes current through the load. With dc supply, the photocell then will exert no further control until switch S1 is momentarily opened to restore control to the gate.

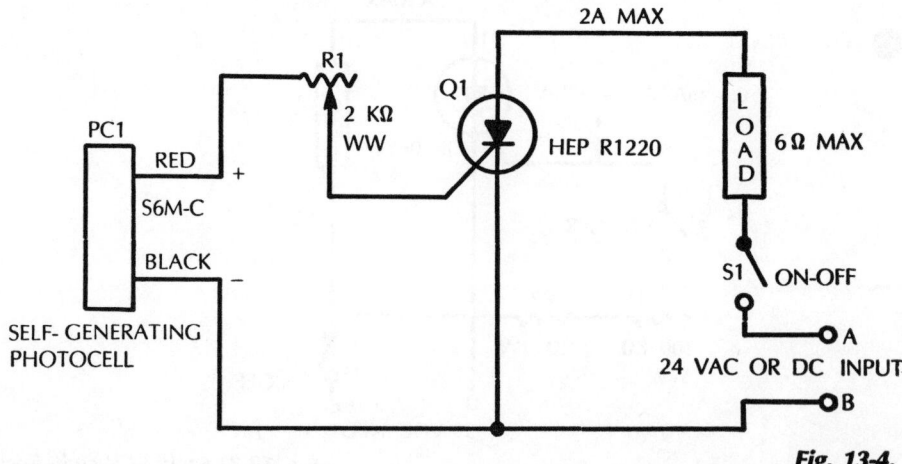

Fig. 13-4.

With ac supply, however, control automatically returns to the gate each time the supply-voltage cycle passes through zero. This means that with ac supply the SCR will switch on each time that light is shined on the cell, and will switch off each time the light is interrupted; but with dc supply, the SCR will switch on each time the cell is illuminated and then will remain on when the light is removed.

This is a relatively simple arrangement, which is useful in many setups requiring a switch that closes on application of a light beam. While the circuit is not so sensitive as some similar ones, the self-generating photocell requires no bias supply, and this further simplifies the circuit. Insertion of the load device into the anode circuit of the SCR obviates the need for an auxiliary electromechanical relay. While a 2-amp SCR operated at 24 volts is shown here, the same circuit may be used with a heavier-current SCR and at higher anode voltage, provided the gate current for the new SCR is within the 8 mA output range of the photocell.

No.125: SCR Light Dimmer

The output of a silicon controlled rectifier can be varied smoothly by varying the phase of the gate trigger, with respect to the anode voltage. The earlier the trigger arrives in the positive half-cycle of anode voltage, the longer the anode current flows and, therefore, the greater its value. Conversely, when the trigger arrives late, anode current flows for a short time and its magnitude is low.

In Fig. 13-5, an adjustable RC-type phase-delay circuit—consisting of R2, R3, and C1—sets the time at which a 2N2646 unijunction transistor (Q2) delivers a gate trigger pulse to switch on the 2N3228 SCR (Q1). (See Chapter 7, for a detailed description of the UJT trigger for SCRs.)

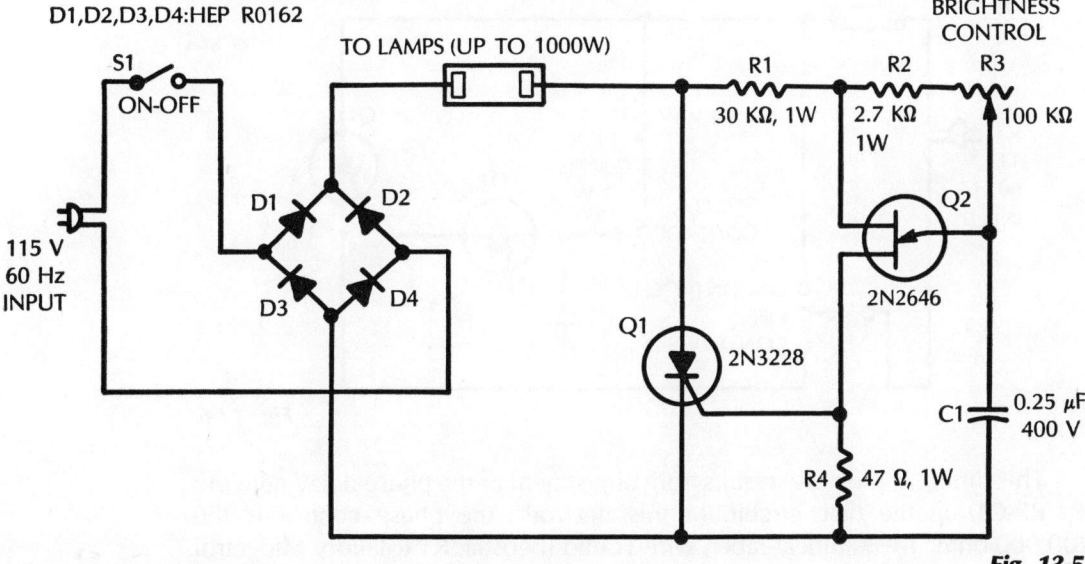

Fig. 13-5.

Through adjustment of the light-duty control, R3, the operator can vary the SCR output over a wide range. In the phase-control network, resistor R2 is a safety device which prevents rheostat R1 from being set to the full anode voltage of the UJT.

This principle is utilized here to control the brightness of incandescent lamps, singly or in groups up to 1000 watts. In this circuit, a full-wave bridge rectifier, consisting of four HEP R0162 silicon power diodes (D1 to D4) supplies rectified power-line voltage to the SCR and lamp load. Because of the full-wave output of the bridge, the SCR is able to handle both half-cycles of the ac line voltage. The phase-shift network is frequency sensitive and has been designed for 60 Hz service (see Table 11-1 for details of this network). The circuit will not work with fluorescent lamps and must not be connected to them.

The 2N3228 SCR shown in Fig. 13-5 is rated at 5-amps, 200-volts, but higher-powered SCRs may be substituted for heavier-duty service. The 2N2646 portion of the circuit may be left intact. In addition to its intended use as a light dimmer, this circuit may be used also as a heater or oven controller.

No. 126: SCR Motor Control

Figure 13-6 shows a motor-speed-control circuit in which the output of a 2N3228 SCR (Q1) is controlled by adjusting the instant during the ac half-cycle at which a gate-trigger pulse arrives from a 1N5411 diac (D1) to trigger the SCR on. The earlier the trigger arrives in the positive half-cycle of SCR anode voltage, the longer the anode current flows until the end of the half-cycle, and the faster the motor runs. Conversely, the later the pulse arrives, the shorter the time of anode-current flow and the slower the motor runs.

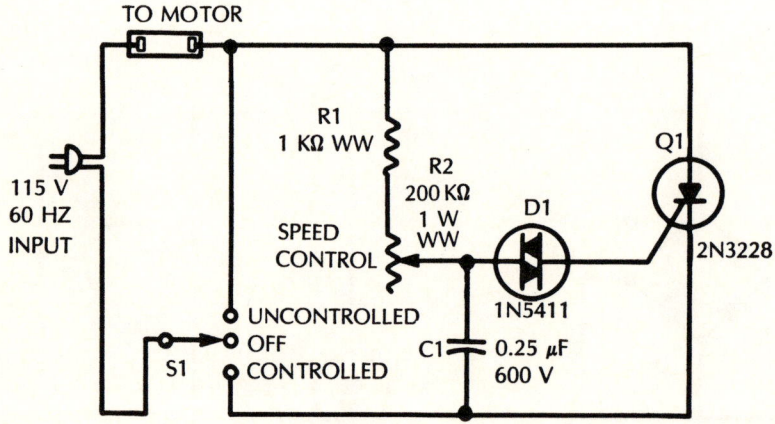

Fig. 13-6.

This timing of the pulse results from adjustment of the phase-delay network (R1-R2-C1) in the diac circuit. In this network, the phase control is the 200,000-ohm, 1-watt, linear-taper, wirewound rheostat, R1 (Mallory Midgetrol

Silicon Controlled Rectifier SCR

or equivalent); and R1 is a safety resistor which prevents R2 from being set to the full anode voltage. (See Chapter 11, for a more detailed description of the phase-controlled-diac type of trigger circuit.) Because the phase-shift network is frequency sensitive, this motor-control circuit is recommended for 60 Hz service only. For operation at other frequencies, other values for R1, R2, and C1 will be required.

Since the SCR in this circuit passes only the positive half-cycles of ac supply voltage, the motor cannot be brought up to full speed. For full-speed operation, throw switch S1 to its uncontrolled position; this connects the motor directly across the power line without benefit of the SCR circuit. With S1 in its controlled position, the speed-control circuit is in operation.

The circuit is useful only with universal motors. With the 2N3228 SCR shown here, motors up to ¾ horsepower can be controlled.

No. 127: Photoelectric Burglar Alarm

Battery-operated burglar alarms are very desirable, since they remain operable during power-line blackouts when house breakins are very likely and in other times of power failure. It is important that the idling current of the device be low, so that one or two hotshot batteries will give long-time service in its operation. Two SCR-type burglar alarm circuits are included in this chapter; one (Fig. 13-7) employs a light beam; the other (Fig. 13-8) uses the familiar arrangement of closed "sensor" switches connected in series.

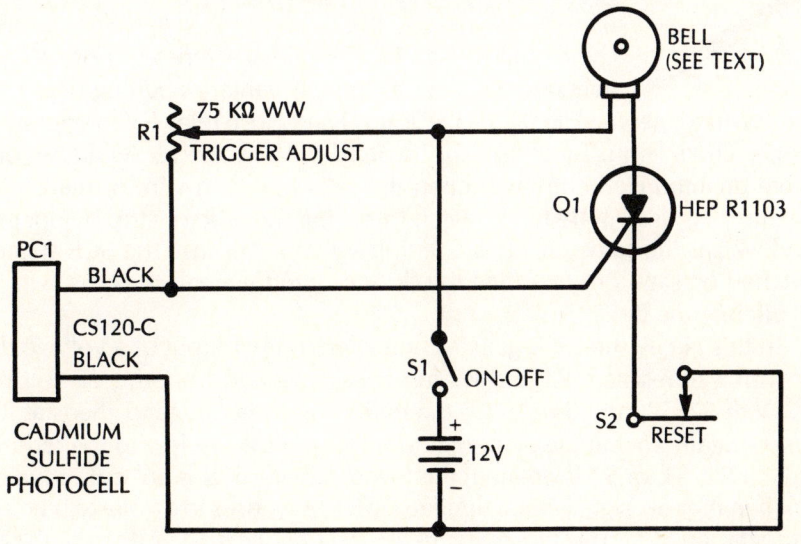

Fig. 13-7.

Figure 13-7 shows a photoelectric-type burglar alarm circuit which goes into operation when a light beam is interrupted, and continues to operate without break after the beam is restored. In this arrangement, a HEP R1103 SCR (A1) is triggered by the dc output of a CS120-C cadmium-sulfide photoconductive cell (PC1). The circuit leg, consisting of resistor R1 and the photocell, constitutes a voltage divider (operated from the 12-volt battery, B1) in which the cell acts as a light-controlled resistor. The output of the voltage divider is applied to the gate of the SCR.

When the cell is illuminated, its resistance is low, and the dc output of the voltage divider then is too low to trigger the SCR into conduction. But when the light beam is interrupted, as by an intruder, the dark resistance of the cell is very high and the output of the R1-PC1 voltage divider rises high enough to trigger the SCR on. The resulting flow of anode current through the bell causes the latter to go into operation. Once the SCR has thus been switched on, the gate loses control, and restoration of the light beam has no effect. The bell may be silenced only by momentarily opening the normally closed switch, S2 (installed at a secret location) which resets the circuit.

To set up the circuit initially, close switch S1 and illuminate the photocell, with R1 set to maximum resistance. Then, darken the cell and adjust R1 to the point at which the bell starts ringing. Depress S2 to silence the bell. This is the correct operating point of the circuit, R1 requiring no subsequent adjustment. Note that the photocell has no exclusive output polarity (see identical color coding of its leads in Fig. 13-7). The bell must be a loud one able to operate efficiently on the SCR anode current (a satisfactory unit is Audiotex 30-9100). Aside from a bell, a siren, horn, or lamp also can be used.

No. 128: Switch-Type Burglar Alarm

The circuit in Fig. 13-8 employs a series of normally closed switches (S2, S3, S4) the momentary opening of any one of which will trigger the HEP R1103 SCR (Q1) and close relay RY to operate a bell, horn, siren, or lamp. These "sensor" switches are installed on windows or doors where unauthorized entry will open the switch. A thin wire or metal-foil strip sometimes is used in lieu of a switch (breaking the wire or strip by opening the window has the same effect as opening switch). Because the SCR when once switched on stays on, reclosing the door or window (and thus the switch) will not silence the bell.

In this circuit, the SCR gate is connected to the output of a voltage divider, consisting of resistors R1 and R2 in series, operated from the 12-volt battery, B1. With S2, S3, and S4 closed, the divider is complete, and rheostat R1 may be set initially so that the dc output of the divider is too low to trigger the SCR. If either S2, S3, or S4 is opened, the lower resistor (R2) is cut out of the divider and then the gate signal, being limited by R1 only, rises to a voltage high enough to trigger the SCR. This closes the relay and connects 12 volts from the battery to the output terminals to actuate the bell or other alarm device. Once the SCR is switched on, it will continue to conduct anode current through the relay—even if the sensor switch is immediately closed—until normally closed switch S5 (installed at a secret location) is momentarily opened to interrupt the anode current and restore control to the gate.

Silicon Controlled Rectifier SCR 207

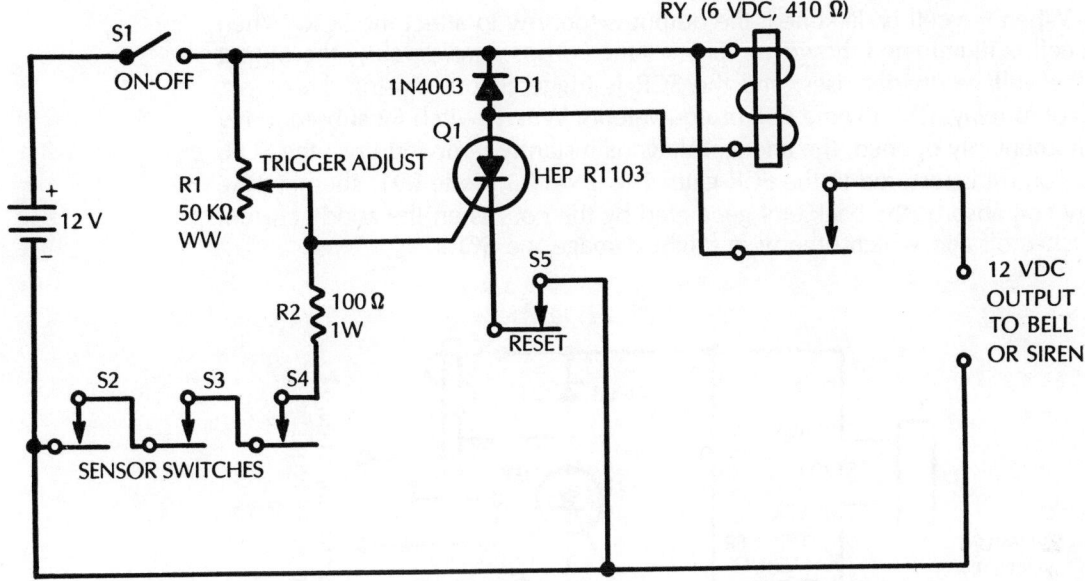

Fig. 13-8. Switch-type burglar alarm.

Relay RY is a 6-volt, 410-ohm dc unit (Sigma 65F1 or equivalent) and has 1-amp contacts which are heavy enough to switch most alarm devices. The 1N4003 diode (D1) shunting the relay coil absorbs the back emf generated by the coil when the anode current switches off and which otherwise might damage the SCR.

No. 129: Photoelectric Garage-Door Opener

Figure 13-9 shows the electronics of a headlight-operated garage-door opener based upon a type HEP R1103 silicon controlled rectifier (Q1). With this setup, illumination of a CS120-C cadmium-sulfide photoconductive cell (PC1) triggers the SCR which, in turn, closes a relay (RY1) to switch ac power to a standard door-opener mechanism. The SCR then remains on until the normally closed switch, S2 (located inside the garage) is momentarily opened, but the disabling switch in the opener mechanism automatically removes the ac power from the mechanism when the door is fully open. To achieve reasonably foolproof service and to avoid accidental triggering by ambient light, the photocell must be provided with a suitable lens system and must be properly hooded so that only the bright light of high beams will actuate the circuit.

In this circuit, the photocell functions as a light-controlled variable resistor exhibiting very high resistance when darkened and low resistance when illuminated. A voltage divider is formed by the photocell and potentiometer R1 in series, and the dc output of this divider is applied to the gate of the SCR.

When the cell is darkened, the output is too low to affect the SCR. When the cell is illuminated, however, its resistance drops considerably, the output of the voltage divider rises, and the SCR is triggered on, closing the 6-volt, 410-ohm relay, RY1 (Sigma 65F or equivalent). When switch S2 subsequently is momentarily opened, the anode voltage is instantly removed from the SCR, and control is restored to the SCR gate. The 1N4003 diode (D1) shunting the relay coil absorbs the back emf generated by the coil when the anode current switches off and which otherwise might damage the SCR.

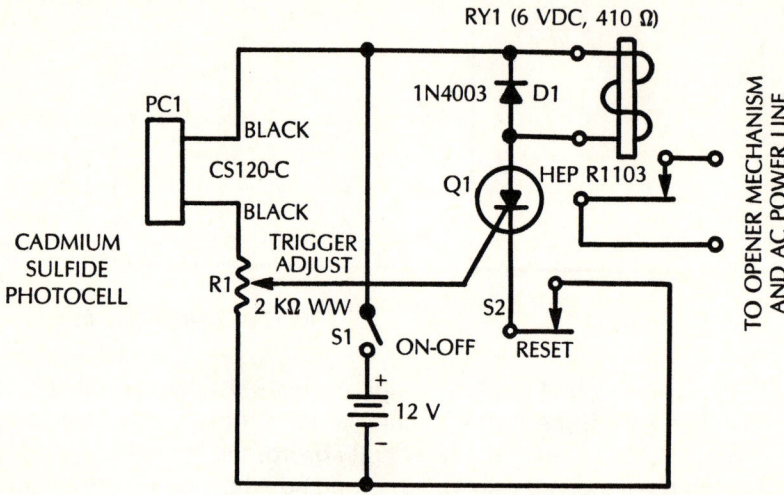

Fig. 13-9. Photoelectric garage-door opener.

To set the circuit initially, darken the photocell, turn potentiometer R1 down to its lowest point, and close switch S1. Next, illuminate the photocell with the headlights that will be used and with these lights spaced from the cell as they normally will be. Then, advance the setting of R1 to the point at which the relay just closes. This is the correct operating point of the circuit, and R1 will require no further adjustment except to compensate from time to time for drift and aging. Finally, darken the photocell again, noting that the relay remains closed. Then, momentarily open switch S2, noting that the relay opens. Note that the photocell has no exclusive polarity; both leads are color-coded black, as shown in Fig. 13-9.

Silicon Controlled Rectifier SCR

No. 130: Variable dc Power Supply

Through phase control of the gate voltage, the dc output of an ac-operated SCR may be varied smoothly over a useful range. Figure 13-10 shows a half-wave rectifier circuit of this kind operated directly from the 115-Vac power line. The average value of the dc output voltage of this circuit can be varied from close to zero to approximately 51 volts by adjustment of the light-duty 100,000-ohm rheostat, R2. The 2N3228 SCR (Q1) is rated at 5-amps, 200-volts. Because this is a half-wave circuit, substantial filtering of the dc output is required, and the resulting dc voltage may be higher than the maximum value given above, depending upon the type of filter used.

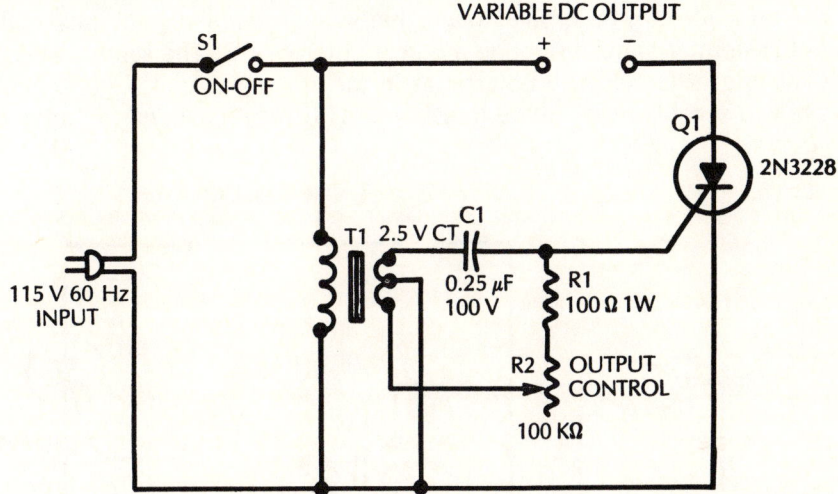

Use a 1:1 isolating transformer when dc + or dc − must be grounded

Fig. 13-10.

The sinusoidal gate signal is derived from the 2.5-volt center-tapped winding of transformer T1 (Stancor P-8629 or equivalent). The phase of this voltage is adjusted by means of an RC-type phase-shift network consisting of resistors R1 and R2 and capacitor C1.

When the gate-trigger value of this voltage is reached early in the anode-voltage positive half-cycle, anode current flows for a longer time (greater angle) during this half-cycle than when the trigger value is reached later. By appropriately adjusting 100,000-ohm rheostat R2, therefore, the operator can choose the phase delay that will result in minimum dc output, maximum dc output, or any point in between. The phase-delay network is frequency sensitive, so the R1, R2, and C1 values given here apply to 60 Hz only. (See Table 11-1, Chapter 11, for performance of this network.)

Because this circuit, as presented in Fig. 13-10, is operated directly from the ac power line, safety precautions must be taken in some of its applications.

For instance, if either the dc+ or dc− terminal is to be grounded, a 1:1 isolating transformer must be inserted between the power line and the input portion (S1-T1) of the circuit, to prevent electric shock or damage to the circuit or to the powered equipment.

The 2N3228 SCR is rated at 5 amps; however, higher-current units may be used in the same circuit for heavier-duty service, provided that their gate-voltage requirements are within the capability of the T1-C1-R1-R2 gate-trigger circuit.

No. 131: High Voltage Variable dc Power Supply

The variable dc power supply described in the preceding section is limited to the ac voltage of the power line. However, the same principle may be applied to a higher-voltage supply by adding a high-voltage transformer and substituting a higher-voltage SCR. The same delayed-phase gate-trigger circuit may be used as in the simpler circuit. Figure 13-11 shows how a suitable high-voltage transformer (T1) may be added to the original circuit.

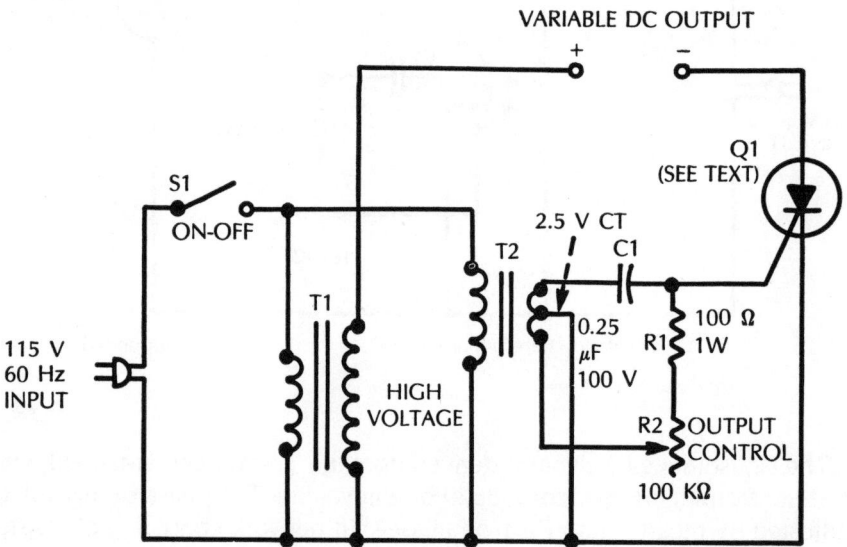

Use a 1:1 isolating transformer when dc + or dc − must be grounded

Fig. 13-11.

In Fig. 13-11, the SCR is still triggered by the low-voltage transformer (T2) and the same phase-shift network (C1-R1-R2), but the anode of the SCR is supplied by the secondary winding of transformer T1. Typical higher voltage SCRs are listed in Table 13-1. The average value of the unfiltered dc output voltage of the high-voltage circuit will be equal to approximately 0.45 times the rms value of the secondary voltage of transformer T1.

Because this circuit, like the earlier one, has a direct connection to the ac power line, special safety precautions must be taken in some of its applications. For instance, to prevent electric shock or damage to the circuit or to the powered equipment, if either the dc+ or dc− terminal is to be grounded, a 1:1 isolating transformer must be inserted between the power line and the input portion (S1-T1-T2) of the circuit.

400 V	2 A.	2N3529
	3.3 A.	40659
	5 A.	2N3525
	7 A.	40379, 40508, 40655, 40657
	10 A.	40739, 40743, 40747
	12.5 A.	2N3670
	20 A.	40751, 40755, 40759
	35 A.	2N3653
600 V	2.5 A.	40813
	5 A.	40640, 40641
	10 A.	40740, 40744, 40748
	16 A.	2N1842A, 2N1850A
	25 A.	2N681, 2N690
	35 A.	2N3873, 2N3899, 40216, 40683, 40735
700 V	2 A.	2N4102
	5 A.	2N4101, 40553, 40555
	12.5 A.	2N4103

Table 13-1. Higher-Voltage SCRs.

No. 132: Dc-To-Ac Inverter

Figure 13-12 shows the circuit of an SCR inverter which runs from a 12-volt battery and delivers 115-volts, 60-Hz ac at 100-watts continuous service and up to 150-watts intermittent service. SCRs give efficient performance in inverters. This circuit employs two 2N3650 SCRs in push-pull, each being triggered by a relaxation oscillator employing a 2N493 unijunction transistor (Q2 and Q3) and their associated frequency-determining networks (R4-R5-C1 and R6-C2). (See Chapter 7, for an explanation of the unijunction relaxation oscillator.)

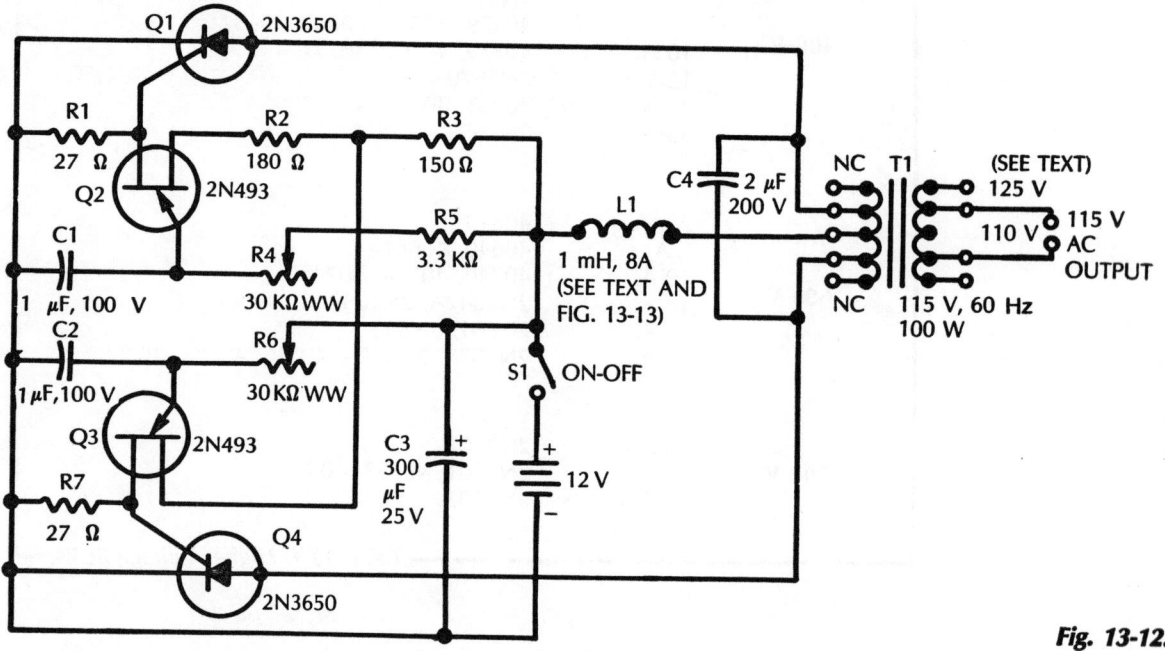

Fig. 13-12.

The 2N3650s are fast-turnoff SCRs recommended especially for inverter service. The upper UJT (Q2) operates at 120 Hz, and the lower one (Q3) at 60 Hz. Once rheostats R4 and R6 are set for these frequencies, they will not ordinarily need readjustment, hence may be provided with slotted shafts for screwdriver adjustment.

In circuits of this type, some means must be included for automatically switching off the SCRs at the proper time. Ordinarily they will continue to conduct, once switched on, and there would consequently be no ac output from the circuit. When this automatic switch-off is accomplished, the SCRs supply pulses alternately to the transformer, T1. This needed commutation is provided by capacitor C4 and inductor L1. As an SCR is switched on, C4 applies a negative voltage momentarily to the anode of the opposite SCR, switching the latter off.

In general, construction of the inverter is straightforward. T1 is a special inverter transformer (Triad TY-75A or equivalent). Choke inductor L1 is a 1-millihenry, 8-amp unit, and since such a heavy-current unit may not easily be located in commercial stocks, winding a simple one should be both economical and time saving. A suitable inductor may be made by closewinding 196 turns of No. 16 enameled wire in 14 layers at 14 turns per layer. Figure 13-13 shows details of a 1¾ in. diameter bobbin for this coil; the bobbin may be turned from wood and later impregnated with insulating varnish—or, if preferred, it can be made from some other dielectric material. The SCRs should be heat-sinked, and the UJTs should be mounted in a cool part of the inverter assembly.

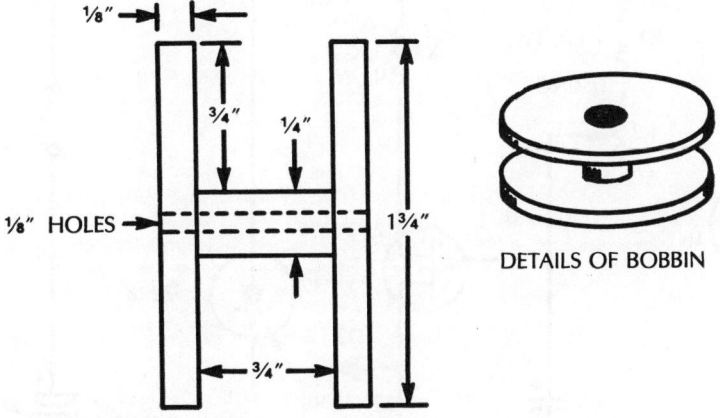

IN BOBBIN, CLOSE WIND 196 TURNS
NO. 16 ENAMELED WIRE (14 TURNS PER
LAYER IN 14 LAYERS)

Fig. 13-13. Details of inverter choke coil.

No. 133: Solid-State Timer

Figure 13-14 shows the circuit of an all-solid-state timer based on a 2N3228 SCR (Q2) which will switch as high as 5 amps through a load device, such as a motor, actuator, heater, lamp, and so on. This is a delayed-make type of timer; that is, the SCR switches on at a selected instant after switch S1 has been closed. Setting of the 500,000-ohm rheostat, R2, allows selection of any interval between 0.1 and 50.1 seconds. While a 28-volt battery is shown here, a well-filtered, power-line-operated supply also can be used.

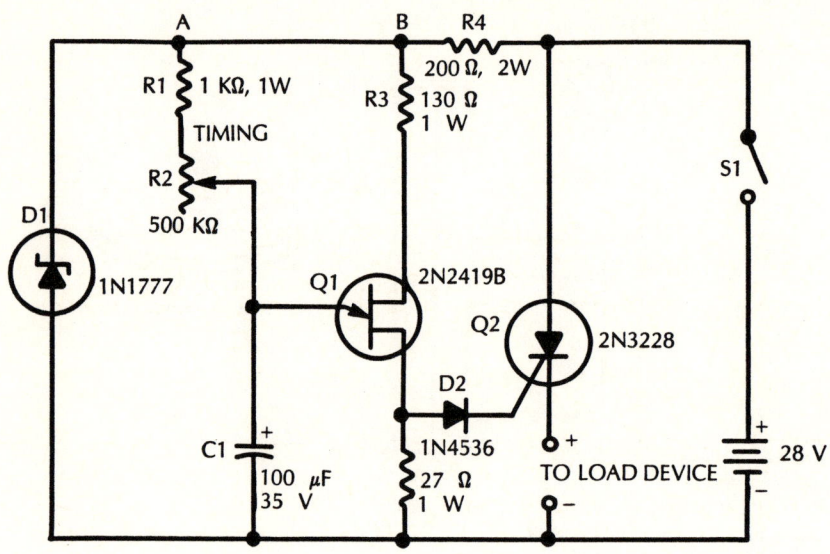

Fig. 13-14.

In this circuit, the SCR is triggered by a pulse from a 2N2419B unijunction transistor (Q1). The anode supply voltage of this UJT, which is also the dc voltage applied to the timing circuit (R1-R2-C1) is regulated by the 1N1777 zener diode (D1) and 200-ohm resistor (R4).

Operation of the circuit is simple: When switch S1 is closed, a regulated +12-Vdc potential is applied to the UJT anode circuit (point B) and the timing circuit (point A). Capacitor C1 then charges through resistors R1 and R2, the voltage across this capacitor increasing according to the time constant of the R1-R2-C1 circuit and thus according to the setting of rheostat R2. When the capacitor voltage eventually reaches the critical value of the UJT, Q1 fires and delivers a positive pulse which passes through steering diode D2 to the gate of the SCR. This triggers the SCR and causes it to conduct current through the load. The SCR then remains on until switch S1 subsequently is opened. Rheostat R2 may be provided with a dial reading directly in seconds on the basis of a calibration of the circuit.

Index

Index

Index

A

ac overload circuit breaker, 182
ac relay, solid-state, 192
active filter
  bandpass, 65
  combination, 64
  conventional bandpass, 68
  high-pass, 63
  low-pass, 61
  notch, 67
active notch filter, 67
AM radio, 74
amplitude-sensitive switch, 177
analog ICs, 51-81
  theory of, 51
audio amplifiers, 52, 55
  two-stage, 15
  with MOSFET input, 57
audio preamplifier, 13
audio thermometer, 106
automatic equipment power switch, 197
automatic FM frequency control, 139
automatic volume limiter, 161

B

bandpass active filter, 65
bar graph display, 166
basic SCR switches, 200
binary counter, 92, 93
binary logic probe, 94
binary numbers, 86
bit shifting, 97
blinking LED, 167
burglar alarm
  photoelectric, 205
  switch-type, 206

C

capacitance measurement, 43
capacitance relay, 45
charge detector, 48
choke coil, 213
circle chase, 97
circuit boards, construction of, 9
circuit breaker
  ac overload, 182
  dc overload, 181
CMOS ICs, 101-117
  handling precautions and, 102
  mixing TTLs and, 102
  theory of, 101
  TTL vs., 101
CMOS logic probe, 105
code-practice oscillator, 71
coil data, 25
coil-winding data, 22
combination active filter, 64
complementary metal oxide semiconductor (CMOS) IC, 82
complex sound generator, 55
components, 3
  handling of, 5
  sources for, 4
  substitution of, 5
continuity checker, 73
continuity testers, 3
controller
  combination thyristor and, 190
  general-purpose, 189
conventional bandpass active filter, 68
cw monitor, 126

D

dc controlled ac lamp, 171
dc overload circuit breaker, 181
dc polarity, 7
dc power supply
  variable, 210
  variable dual, 195
dc voltage standard, 162
dc-controlled solid-state ac relay, 192
dc-controlled triac switch, 191
dc-equipment protector, 162
dc-to-ac inverter, 212
decimal counter, 168
diacs, 175-184
  theory of, 175
dice, electronic, 95
DIGI-KEY Corporation, 2
digital circuits, 86
digital ICs, 82-100
  theory of, 82
digital logic, 83
direct-reading capacitance meter, 27
  initial calibration of, 29
documentation, 10
dual LED flasher, 72

E

egg timer, 107
18-volt, 1 regulated dc supply, 155
electrically latched relay, 179
electronic dc millivoltmeter, 60
electronic dc voltmeter, 25
  initial calibration for, 27
electronic dice, 95
electronic noise maker, 77
electroscope, 48
emitting diodes, 164-174
  theory of, 164
enclosures, construction of, 9
equipment power switch, 197
exponentiation, 97
external magnetic fields, 7

F

fiberoptic infrared link, 174
field-effect transistors (FETs), 11-30
  junction-type action of, 13
  schematic of, 12
  theory of, 11
filter
  active notch, 67
  bandpass active, 65
  combination active, 64
  conventional bandpass active, 68
  high-pass active, 63
  low-pass active, 61
5-volt, 1.25 regulated dc supply, 153
555 and 556 timers, 54
flasher, dual LED, 72
flasher IC, 3903, 54
FM frequency control, automatic, 139
free-running multivibrator, 122
frequency counter, 3, 113
  calibration of, 113
  reading of, 114
frequency doubler, 143
frequency modulators, 140
frequency multipliers, 143
frequency tripler, 144
full adder, truth table for, 91

G

garage-door opener, photoelectric, 208
gates, 85, 86
general-purpose controller, 189
general-purpose rf amplifier, 34
grounds, 6

H

half adder, 89
high-pass active filter, 63
higher voltage dc regulators, 150
higher-voltage SCRs, 211
hot case, 6

I

IC extractor, 2
IC inserter/pin straightener, 2
IC sockets, construction of, 9
inductance measurement, 43
infrared-emitting diode detection, 166
infrared-emitting diode transmissions, 166
infrared link, 173
interval timer, 44

218 Index

inverter, dc-to-ac, 212
isolation, 6

J

junction type field-effect transistor (JFET), 11

L

latched relay, 179
latching sensor circuit, 180
leads, 6
light dimmer, 194
   SCR, 203
light meter, 172
light-controlled SCR, 202
light-duty regulated dc supply, 152
light-emitting diodes (LEDs), 164
   blinking, 167
   three-lead tricolor, 165
   two-lead tricolor, 165
logic gates, 85, 86
logic operations, 85, 86
logic probe, 3
   binary, 94
   CMOS, 105
   tricolor, 167
low-pass active filter, 61
low-resistance dc milliammeter, 59
LSI counter, 93

M

magnetic fields, external, 7
magnitude comparator, 87
melody maker, 99
metal oxide semiconductor field-effect transistor (MOSFET), 31-50
   theory of, 31
metronome, 111, 128
milliammeter, low-resistance dc, 59
millivoltmeter, 60
MOSFET, 31
motor controls, 193
   SCR, 204
multiple-output dc voltage regulator, 151
multiplication, 97
multivibrator
   free-running, 122
   one-shot, 123
music box, 115

N

NAND functions, 85
night light, 76
noise maker, electronic, 77
NOR functions, 85
NOT functions, 85
notch filter, active, 67

O

one-shot multivibrator, 123
op amps, 53
OR functions, 85
oscillator
   code-practice, 71
   frequency modulated self-excited, 141
   relaxation, 124
   standard-frequency, 125
   voltage-tuned rf, 138
oscilloscope, 3

P

pendulum clock, 103
phase angles, 184
phase-controlled trigger circuit, 183
phase-shift audio oscillator, 18
photoelectric burglar alarm, 205
photoelectric garage-door opener, 208
power supply
   connection of, 6
   type of, 7
   variable dc, 210
   variable dual dc, 195
power switch, automatic, 197
product detector, 19
projects
   18-volt, 1 regulated dc supply, 155
   2 1/2-watt intercom, 57
   5-volt, 1.25 regulated dc supply, 153
   ac overload circuit breaker, 182
   active notch filter, 67
   AM radio, 74
   amplitude-sensitive switch, 177
   audio amplifier, 55
   audio amplifier with MOSFET input, 57
   audio preamplifier, 13
   audio thermometer, 106
   automatic equipment power switch, 197
   automatic FM frequency control, 139
   automatic volume limiter, 161
   bandpass active filter, 65
   bar graph display, 166
   basic SCR switches, 200
   binary counter, 92
   binary logic probe, 94
   blinking LED, 167
   capacitance relay, 45
   circle chase, 97
   CMOS logic probe, 105
   code-practice oscillator, 71
   combination active filter, 64
   combination thyristor/controller, 190
   continuity checker, 73
   conventional bandpass active filter, 68
   cw monitor, 126
   dc overload circuit breaker, 181
   dc voltage standard, 162
   dc-controlled ac lamp, 171
   dc-controlled solid-state ac relay, 192
   dc-controlled triac switch, 191
   dc-equipment protector, 162
   dc-to-ac inverter, 212
   decimal counter, 168
   direct-reading capacitance meter, 27
   dual LED flasher, 72
   egg timer, 107
   electrically latched relay, 179
   electronic dc millivoltmeter, 60
   electronic dc voltmeter, 25
   electronic dice, 95
   electronic noise maker, 77
   electroscope, 48
   fiberoptic infrared link, 174
   free-running multivibrator, 122
   frequency counter, 113
   frequency doubler, 143
   frequency fluctuating quartz crystals, 142
   frequency modulated self-excited oscillators, 141
   frequency tripler, 144
   general purpose controller, 189
   general purpose rf amplifier, 34
   half adder, 89
   high-pass active filter, 63
   higher voltage dc regulators, 150
   infrared emitting diode detection, 166
   infrared link, 173
   infrared-emitting diode transmissions, 166
   interval timer, 44
   latching sensor circuit, 180
   light dimmers, 194
   light meter, 172
   light-controlled SCR, 202
   light-duty regulated dc supply, 152
   low-pass active filter, 61
   low-resistance dc milliammeter, 59
   LSI counter, 93
   magnitude comparator, 87
   melody maker, 99
   metronome, 111, 128
   motor controls, 193
   multiple-output dc voltage regulator, 151
   music box, 115
   night light, 76
   one-shot multivibrator, 123
   pendulum clock, 103
   phase-controlled trigger circuit, 183
   phase-shift audio oscillator, 18
   photoelectric burglar alarm, 205

Index

photoelectric garage-door opener, 208
product detector, 19
proximity detector, 75
pulse and timing generator, 121
pulse generator, 120
Q multiplier, 41
random number generator, 107
regenerative receiver, 21
regulated voltage divider, 157
relaxation oscillator, 124
remotely controlled tuned circuit, 137
rf harmonic intensifier, 145
SCR light dimmer, 203
SCR motor control, 204
seven-segment displays, 166
signal tracer, 69
simple ac voltage regulator, 160
simple dc voltage regulator, 149
simple triac switch, 187
simple voltmeter, 169
siren, 110
solid state timer, 214
sound machine, 79
source follower, 39
standard-frequency oscillator, 125
static dc switch, 178
superregenerative receiver, 23
switch-type burglar alarm, 206
ten-meter preamplifier, 36
three-lead tricolor LED, 165
tone-identified signal system, 129
touch-plate relay, 46
transistor-bias regulator, 158
tricolor logic probe, 167
trigger for SCR, 130
tuned crystal oscillator, 17
two-lead tricolor LED, 165
two-stage audio amplifier, 15
ultrasonic pickup, 49
untuned crystal oscillator, 16
variable dc power supply, 210
variable dual dc power supply, 195
VCD-tuned LC circuits, 135
voltage regulator for tube heater, 159
voltage-regulated dual dc supply, 156
voltage-tuned rf oscillator, 138
voltage-variable capacitor, 133
wide-range LC checker, 42
wideband instrument amplifier, 38
proximity detector, 75
pulse and timing generator, 121
pulse generator, 120

Q
Q multiplier, 41

R
random number generator, 107
regenerative receiver, 21
regulated voltage divider, 157
relaxation oscillator, 124
remotely controlled tuned circuit, 137
rf amplifier, 34
rf harmonic intensifier, 145

S
safety, 3
SCR light dimmer, 203
SCR motor control, 204
SCR switches, basic, 200
semiconductors, handling of, 6
sensitive dc-controlled triac switch, 191
seven-segment displays, 166
signal generators, 3
signal tracer, 69
silicon controlled rectifier (SCRs), 198-214
  higher-voltage, 211
  light dimmer using, 203
  light-controlled, 202
  theory of, 198
  trigger for, 130
simple ac voltage regulator, 160
simple dc voltage regulator, 149
simple triac switch, 187
simple voltmeter, 169
siren, 110
soldering, 6
soldering iron, 2
solid state timer, 214
sound generator, complex, 55
sound machine, 79
source follower, 39
standard-frequency oscillator, 125
static dc switch, 178
static discharge, 102
superregenerative receiver, 23
switch-type burglar alarm, 206

T
ten-meter preamplifier, 36
test equipment, 3
thermometer, audio, 106
three-lead tricolor LED, 165
3903 flasher, 54
thyristor, 190
timers
  555 and 556, 54
  solid state, 214
tone-identified signal system, 129
tools, 1
  hand, 2
touch-plate relay, 46
transistor-bias regulator, 158
transistor-transistor logic (TTL) IC, 82
  CMOS ICs vs., 101
  mixing CMOS ICs and, 102
triac switch, simple, 187
triacs, 185-197
  theory of, 185
tricolor logic probe, 167
trigger circuit, phase-controlled, 183
trigger for SCR, 130
troubleshooting, 7
  breadboard, 7
  working projects, 8
truth tables, 84
tuned crystal oscillator, 17
tuning capacitance, 25
2 ½-watt intercom, 57
two-lead tricolor LED, 165
two-stage audio amplifier, 15

U
ultrasonic pickup, 49
unijunction transistor (UJT), 118-130
  theory of, 118
untuned crystal oscillator, 16

V
variable dc power supply, 210
variable dual dc power supply, 195
variable-capacitance diode (VCD), 131-146
  quartz crystal using, 142
  self-excited oscillators using, 141
  theory of, 131
VCD-tuned LC circuits, 135
voltage regulator for tube heater, 159
voltage-regulated dual dc supply, 156
voltage-tuned rf oscillator, 138
voltage-variable capacitor, 133
voltages, specified, 7
voltmeter, simple, 169
volume limiter, automatic, 161

W
wide-range LC checker, 42
  capacitance measurement with, 43
  inductance measurement with, 43
wideband instrument amplifier, 38
wiring, 6
work area, 2

X
X-Acto knife, 2

Z
zener diodes, 147-163
  theory of, 147

Other Bestsellers From TAB

☐ **44 POWER SUPPLIES FOR YOUR ELECTRONIC PROJECTS**—Robert J. Traister and Jonathan L. Mayo

Here's a sourcebook that will make an invaluable addition to your electronics bookshelf whether you're a beginning hobbyist looking for a practical introduction to power supply technology, with specific applications . . . or a technician in need of a quick reference to power supply circuitry. You'll find guidance in building 44 supply circuits as well as how to use breadboards, boards, or even printed circuits of your own design. You can build a full-wave bridge—supply a dual voltage power supply—a 9-volt series-regulated supply—a 28-volt power supply using three zener diodes method for alternating secondary voltage—a higher current 5-volt regulated supply switching regulator supply—*and many more!* 220 pp., 208 illus.

Paper $15.95 **Hard $24.95**
Book No. 2922

☐ **COMPUTER ARCHITECTURE AND COMMUNICATIONS**—Neil Willis

This indispensable guide presents in a logical and straightforward way, the ideas and concepts of computer architecture and the principles of computer communications. It takes you inside your computer to see what goes on and how it's accomplished. Covering the full range of computer architecture, it will serve as your friendly guide through every part of your computer . . . from the building blocks that go to make up a computer to computer networks. 288 pp., 127 illus. 6″ × 9″

Paper $16.95 **Hard $24.95**
Book No. 2870

☐ **FIBEROPTICS—A REVOLUTION IN COMMUNICATIONS—2nd Edition**—John A. Kuecken

Get an up-to-the-minute overview of the hottest new technology to hit the communications industry in decades! Aimed at providing a working knowledge of fiberoptic devices, this comprehensive sourcebook takes you from the basics of why and how fiberoptics were invented, right through how they work and their applications to almost any electronic purpose. You'll find dozens of practical fiberoptic applications . . . everything from voltage-to-frequency converters, measurements in strong electrical fields, and computer data linkage to mechanical isolation and local area networks. 352 pp., 166 illus.

Hard $28.95 **Book No. 2786**

☐ **LASERS—THE LIGHT FANTASTIC—2nd Edition** Clayton L. Hallmark and Delton T. Horn

An up-to-date course on laser development, theory, hardware and applications. If you want to experiment with lasers, you will find the guidance you need here—including safety techniques, a complete glossary of technical terms, actual schematics, and information on obtaining the necessary materials. There's even a laser project that you can build at home. 280 pp., 129 illus.

Paper $12.95 **Hard $19.95**
Book No. 2905

☐ **POWER CONTROL WITH SOLID-STATE DEVICES**—Irving M. Gottlieb

Whether you're an engineer, technician, advanced experimenter, radio amateur, electronics hobbyist, or involved in any way in today's electronics practice, you'll find yourself turning to this book again and again as a quick reference and as a ready source of circuit ideas. Author Irving Gottlieb, a professional engineer involved in power engineering and electronic circuit design, examines both basic and state-of-the-art power control devices. 384 pp., 235 illus., 6″ × 9″

Hard $29.95 **Book No. 2795**

☐ **ELEMENTARY ELECTRICITY AND ELECTRONICS —Component by Component**—Mannie Horowitz

Here's a comprehensive overview of fundamental electronics principles using specific components to illustrate and explain each concept. This approach is particularly effective because it allows you to easily learn component symbols and how to read schematic diagrams as you master theoretical concepts. You'll be led, step-by-step, through electronic components and their circuit applications. Horowitz has also included an introduction to digital electronics. 350 pp., 231 illus.

Paper $16.95 **Hard $23.95**
Book No. 2753

Other Bestsellers From TAB

☐ **DIGITAL ELECTRONICS TROUBLESHOOTING—2nd Edition—Joseph J. Carr**

Now, the Electronics Book Club brings you a brand new, completely updated, and expanded edition of this classic guide to digital electronics troubleshooting. It covers not only the basics of digital circuitry found in the first edition, it also provides details on several forms of clock circuits, including oscillators and multivibrators, and up-to-the-minute coverage of microprocessors used in today's cassette players, VCRs, TV sets, auto fuel and ignition systems, and many other consumer products. Adding to its effectiveness are more than 300 diagrams, schematics, and charts clearly illustrating each concept. 420 pp., 346 illus.

Paper $17.95   **Hard $25.95**
Book No. 2750

☐ **DESIGN AND APPLICATION OF LINEAR COMPUTATIONAL CIRCUITS—George L. Batten, Jr.**

In this easy-to-follow guide, Batten shares his tested and proven methods of constructing and using analog circuitry. Equipment needed to build the circuits, including the needed power supplies, is covered and there's a complete analysis of computational circuits. Finally, you learn how to program these circuits to solve differential equations, to generate other functions, and for elementary process control applications. 208 pp., 136 illus., 6" × 9" with dust jacket

Hard $24.95   **Book No. 2727**

☐ **UNDERSTANDING OSCILLATORS—2nd Edition—Irving M. Gottlieb**

At last, here's a guide to designing, operating, and using oscillators that presents a clear picture of the relationship between the theory of oscillators and their practical use. Completely up-to-date on the latest oscillator technology, including solid-state oscillator circuits, it provides you with the information you need to devise oscillators with optimized performance features, to service systems highly dependent upon oscillator behavior, and to understanding the many trade-offs involved in tailoring practical oscillators to specific demands. 224 pp., 157 illus.

Paper $12.95   **Hard $16.95**
Book No. 2715

☐ **30 CUSTOMIZED MICROPROCESSOR PROJECTS—Delton T. Horn**

Here it is! The electronics project guide that you've been asking for—a complete sourcebook on designing and building special purpose computer devices around the Z80 microprocessor! Includes building instructions, detailed schematics, and application programs for 30 intriguing and useful dedicated CPU projects, plus how to customize these devices for your own individual applications! 322 pp., 211 illus.

Paper $14.95   **Hard $22.95**
Book No. 2705

*Prices subject to change without notice.

Look for these and other TAB books at your local bookstore.

TAB BOOKS Inc.
P.O. Box 40
Blue Ridge Summit, PA 17214

**Send for FREE TAB catalog describing over 1200 current titles in print.
Or Call For Immediate Service 1-800-233-1128**